THE ESSENTI<
V E D
MATHEMA

**Rajesh Kumar Thakur** currently works as a director at the National Vedic Maths Academy, a branch of the All India Ramanujan Maths Club, Gujarat. He has postgraduate degrees in three diverse subjects: mathematics, operations research and education. He has been teaching secondary and senior secondary school students for the past twelve years and has written over twenty books and around 100 articles on mathematics. To popularize Vedic mathematics, he has conducted over 180 nationwide seminars and trained more than 40,000 students. He has conducted several talks on All India Radio and at the National Level Maths Quiz Show. He has also written many poems, dozens of which have been published in various magazines of repute. Thakur has received many awards, including the National Best Teacher Award from AlRMC in 2010.

# THE ESSENTIALS OF
# VEDIC
# MATHEMATICS

Rajesh Kumar Thakur

RUPA

Published by
Rupa Publications India Pvt. Ltd. 2013
7/16, Ansari Road, Daryaganj
New Delhi 110002

*Sales centres:*

Allahabad Bengaluru Chennai
Hyderabad Jaipur Kathmandu
Kolkata Mumbai

ISBN: 978-81-291-2374-9

10 9 8 7 6 5 4 3 2 1

The moral right of the author has been asserted.

Typeset in Caxton BK BT 10/14

Printed at Replika Press Pvt. Ltd., Haryana

*To My Son*
*Nilabh Thakur*

# Contents

Foreword                                    *xi-xii*
Preface                                     *xiii-xiv*
*What is Vedic Mathematics?*                *xv–xvii*
*Why is Vedic Mathematics Essential?*       *xix–xxi*
*Vedic Sutras*                              *xxiii–xxiv*
*Vedic Sub-sutra*                           *xxv–xxvi*

1.  **Addition:** Introduction, Vedic sutras for addition – *Purna puranabhyam, Sankalan Vyavkalanabhyam* and *Ekadhikena Purvena.* Meaning of Vedic Sutras and their applications.            1–8

2.  **Subtraction:** Introduction, Vedic Sutra for subtraction – *Nikhilam Navatascaramam Dastatah and Vinculum.* Meaning and application, *Digit Separator Method* for subtraction.            9–24

3.  **Multiplication:** Introduction, Vedic sutras:- *Antyayor dasakepi, Nikhilam Navatascaramam Dastatah, Anurupyena, Ekanyuena Purvena , Antyayoshatakepi, Vamanlvavoh Dasake api, Vamanlyayoh Dasake Gunijah Api and Urdhva-Tiryak,* their meanaings and applications. *Dot and Stick Method* of Multiplication, Multiplication of three numbers, Multiplication of four numbers.            25–77

4. **Multiplication through Observation:** 78–91
   Introduction, Multiplication of a number
   through 11, 111, 1111, 25, 125, 625, 5, 50,
   51 etc.

5. **Multiplication in Algebra:** Introduction, 92–102
   Urdhva Tiryagbyama sutra of multiplication
   of Binomial and Trinomials. Multiplication of
   polynomials having unequal number of terms.

6. **Division:** Introduction, Vedic sutra- *Nikhilam,* 103–127
   *Paravartya Yojayet, Urdhva Tiryag and
   Dhwajanka method* – their meanings and
   applications.

7. **Square:** Introduction, Vedic Method for 128–142
   squaring: *Ekadhikena Purvena, Yavadunam
   Tavduni kritya Vargena Yojayet, Duplex method
   and Urdhva Tiryagbhyam,* its meaning and
   applications.

8. **Square Root:** Introduction, *Vilokanam and* 143–157
   *Duplex Vedic Method of finding square root.*

9. **Square Root of Irrational Number:** Introduction, 158–160
   Vedic method Vilokanam and use of Differential
   Calculus in extracting square root.

10. **Cube:** Introduction, Vedic Sutras: *Yavadunam,* 161–171
    *Anurupyena and Nikhilam,* theirs meanings
    and applications.

11. **Cube Root:** Introduction, Vedic Method 172–185
    Vilokanam and Beejank for finding the cube
    root of any number.

12. **Fourth Power of a Number:** Introduction, 186–195
    Pascal Triangle, the method of finding the
    fourth power of a number.

13. **Fourth Root of a Number:** Introduction, Vedic    196–202
    sutra –Vilokanam, its meaning and application.

14. **Simultaneous Equation:** Introduction, Vedic    203–211
    Method – Paravartya Yojayet, Anurupye Sunyam
    anyat, Sankalana- Vyavkalana bhyam, theirts
    meanings and applications.

15. **Cubic Factorization:** Introduction, Vedic method    212–217
    – Gunit Samucchaye Samuccaye Gunita, their
    meanings and applications.

16. **Quadratic Equation:** Introduction, different    218–227
    Vedic sutras- Vilokanam, *Sunyam Samya
    samuccaya, Anurupye and Sunyam anyat* –
    their meanings and applications.

17. **Casting Out Nines:** Introduction, Fundamental    228–239
    Rule of Navasesh, Vedic method of checking
    the accuracy of addition, subtraction,
    multiplication, division, square, and cube.

18. **Trigonometry:** Introduction, How to compute    240–249
    Pythagorean Triplets, Computing trigonometric
    ratio of Sin A, Sin 2A, Sin 3A, Sin A/2  Sum
    and Difference of Compound Angles, etc.

19. **Questions for Practice**    250–260

20. **Feedback from Students**    261–263

21. **Bibliography**    265–667

# Foreword

Vedic mathematics is now a popular name amongst students preparing for competitive examinations like CAT, XAT, Olympiad, and Engineering etc. The brainchild of Sri Bharti Krishna Tirtha Ji Maharaj, Vedic mathematics has scaled heights since its inception. It is being taught in some of the most prestigious institutions in England, America and Australia. NASA scientists are said to have applied its principles in the area of artificial intelligence. Swamiji himself has impressed the world with his unique techniques and in a talk in 1958 in America, Sri Krishnaji said:– 'People who have practical knowledge of the application of the sutras, need not go in for the theory side of it all.'

The importance of Vedic mathematics can be understood with the fact that it has been adopted in the curriculums of secondary boards in states like Rajasthan, Gujrat and Maharastra. Moreover, medical research has proved that Vedic mathematics keeps both sides of our brain fit. In cut-throat competition where time is the only constraint, we need to train our brain in such a way that it is able to do fast calculations in a fraction of a minute, and that too without using pen and pencil. Vedic mathematics, in my opinion, does the same. It helps in saving time almost one-tenth of what we do in the traditional approach of solving a problem. It has wider application and even subjects like Geometry, Calculus, Algebra, Arithmetic, and Trigonometry can be handled well with the Vedic techniques.

A recent survey of NCERT shows that more than 50 % of students passing their primary school examination are unable to do simple arithmetical calculations. The reason behind it is the old mathematical technique being followed by schools for centuries. I personally feel that fear of mathematics can be removed if the application of Vedic mathematics is taught to students from the primary level onwards.

The present book written by Mr Rajesh Thakur will not only help students preparing for competitive examinations, but it will also serve the needs of students who want to excel in their Board examination as the syllabus covered in the secondary level in different state boards has also been included. I have witnessed the technique of Mr Rajesh Thakur and his unique style of presenting mathematical problems in the simplest way. This made me feel that Vedic mathematics should be made a compulsory subject in India.

This book is a ready reference as the presentation of the book makes it easy to learn. The language of the book is very simple, and ample examples given in each chapter will make the understanding of the concept easy. The manner in which this book has been written to cater the need of even a lay man, will make it popular in the market. I am hopeful that this book will bring a change in the thinking style of students and they will enjoy doing mathematical calculations. A day will come when India will again rule the world and prove to be a spiritual and Vedic guru to the world. I congratulate Mr Rajesh Thakur and wish the publisher of the book all the very best for the success of this book.

Dr J.J. Rawal
President
Indian Planetary Society, Mumbai

# Preface

The noted mathematician Carl Fredric Gauss writes, 'Mathematics is the queen of all subjects.' This quotation can't be said to be 100% true in the present scenario, as a recent survey by NCERT says that more than 50% of class V students don't know how to do simple arithmetical operations like addition and subtraction. The Annual Status of Education (ASER) 2011 pointed out that the situation is something more alarming. The ASER report produced by a coalition of non-government organizations has found that less than a third of class III students in rural Indian schools can solve two-digit subtraction problems.

The situation is so because the traditional method has no room for the practical method. The methods of addition, subtraction, multiplication and division taught in our primary schools are lengthy and time-consuming. The methods don't allow counter-checking in-between to confirm whether the result obtained is right or wrong. Vedic mathematics brings a change in the attitude of students as there are ample methods of solving a single sum and the student is the best judge of which method to use in a particular situation. Moreover, the operation is so simple that hardly anyone falls in the wrong line. The best part of Vedic mathematics is that you can check your calculation and know whether you are right or wrong in a few seconds.

The sixteen Vedic sutras open a new horizon for students and these sutras can allow you to practice the eight fundamental arithmetic operations (Addition, Subtraction, Multiplication,

Division, Square, Square root, Cube and Cube root) in a few seconds. This method helps you to calculate fast in your class room, school exams, and also help you to compete in competitive examinations as it saves almost one-tenth of the time taken by the traditional long method.

I have been teaching Vedic mathematics since the past ten years, conducted 180 workshops in ten states and trained around 40,000 students. I have discovered that even a common man with little knowledge of mathematics enjoys learning the Vedic way of calculation.

I have tried to include the maximum number of topics in this book to cater to the need of students preparing for competitive examinations. The presentation of the book is simple and self-explanatory. I hope the reader will love to learn the Vedic method of calculation and enjoy it.

I would like to extend my thanks to my parents who always supported me in popularizing this ancient technique amongst the masses. I can't forget the tremendous effort of my wife, brothers and friend, Dr Piyush Kumar, who encouraged me to write this book and provided every help they could in completing it.

Despite all efforts, however, some errors might have crept in. I shall be grateful to all my readers if the same are brought to my notice. Any suggestion towards improvement of the book will be gratefully acknowledged.

Rajesh Kumar Thakur
E-mail: rkthakur1974@gmail.com

# What is Vedic Mathematics?

यथा शिखा मयूराणां नागानां मणयो यथा ।
तद्धद्धेङ्ग-शास्त्राणां, गणितं मूर्धानि स्थितम् ।।

(Like the crest on the heads of peacocks, like the gems on the hoods of cobras, mathematics is at the top of Vedanga Jyotisha.)

Vedic Mathematics has its origin in the Vedas. The word Veda literally means the 'fountain head and illimitable storehouse of all knowledge.' Therefore, the Vedas contain all knowledge that is essential for mankind. Vedic mathematics is a collection of sixteen beautiful formulae from the Vedas, discovered by His Holiness, Jagadguru Sankaracharya Sri Bharati Krishna Tirthaji Maharaj. It is a gift to the world by Swamiji, who himself was a great scholar. The original source of Vedic mathematics is the Atharva Veda and all the sutras and sub-sutras were rediscovered by Swamiji between 1911 and 1918. Though you can't find all the sixteen formulae in the Vedas, as they don't appear in them,it is the ultimate discovery of Swamiji, after extensive research of the Vedas and Upaveda. It is not important for us to know whether the origin of these sixteen sutras is from the Vedas or not, as long as it is handy in saving our precious time. As our scriptures say:-

युक्तियुक्तं वचो ग्राह्यं बालादपि शुकादपि ।
युक्तिहीनं वचस्त्याज्यं वृद्धादपि शुकादपि ।।

( Whatever is consistent with right reasoning should be accepted, even if it comes from a boy, or even from a parrot; and whatever is inconsistent therewith ought to be rejected, whether emanating from an old man, or even from the great sage, Shree Shuka himself.)

Vedic mathematics is the name of the wind that has created revolutionary changes in fast calculations. It is a super-fast way of making all mathematical calculations easier and faster than the traditional one. Nowadays, it has become a must-learn tool for students who want to perform faster and flawless calculations in a few seconds. Students preparing for CAT, XAT, UPSC, SSC, NTSE, and banking exams have now started adopting the Vedic mathematics technique in solving arithmetical problems involving addition, subtraction, multiplication, division, and square, square root, cube, cube root etc., thus saving a lot of time.

All the sixteen formulae deal with the different branches of mathematics. The sutras being one-line phrases, are easy to understand and remember. They speak for the subject's coherence and simplicity in handling mathematical problems. These sixteen formulae can be used to solve problems ranging from arithmetic to algebra and geometry to calculus and trigonometry. The importance of Vedic mathematics can be understood by the fact that complex mathematical questions which otherwise take numerous steps to solve can be solved through Vedic mathematics mentally or in only a few seconds. These sutras are so beautifully interrelated that a single formula can be used to perform different arithmetical operations. Moreover, the beauty of Vedic mathematics can be judged by the fact that it provides ample room for experimentation. All the fundamental operations can be performed by various methods and the learner has a choice to use the method he/she feels comfort with. Even a layman after being equipped with the Vedic Method can multiply any 5 x 5 digit multiplication in less than 20 seconds and that too, in a single line. This process boosts confidence in one's ability and keeps fear away. Vedic mathematics is an important tool for all because:

- It is truly unconventional and lucid in its methodology.
- Within second it helps to carry out tedious and

cumbersome mathematical calculations from different branches of mathematics.

- It converts dry and insipid maths into a playful and joyful activity.
- It reduces the "maths-phobia" many suffer from.
- The accuracy of a problem can be judged in seconds.
- It is ten times faster than the traditional method taught in our classrooms.
- In the Vedic system, difficult problems can often be solved very quickly, and calculations can be carried out mentally involving minimal paper calculations.

# Why is Vedic Mathematics Essential?

Vedic mathematics is an ancient technique consisting of sixteen sutras and sixteen sub-sutras, which simplifies not only the eight fundamental arithmetical operations, but also handles algebraic concepts like simultaneous equation, simple equation, quadratic equation, factorization of cubic equations etc., more effectively than the traditional approach. Apart from this, Vedic mathematics works faster in areas of Trigonometry, Co-ordinate Geometry and Calculus, in the same effective manner. In fact, there is no part of mathematics, pure or applied, that is beyond its jurisdiction. Dr LM Singhvi, former High Commissioner of India in the UK writes – 'A single sutra would generally encompass a varied and wide range of particular applications and may be likened to a programmed chip of our computer age.'

## What is the speciality of Vedic Mathematics?

Vedic mathematics helps students in minimising careless mistakes. It is simple and a one-line approach. Moreover, it has an inbuilt system of a series of checks. The danger-prone areas like addition, subtraction, multiplication and division can be checked in a fraction of minutes by using the Vedic technique. Even cubic factorization, square, square root, quadratic equation and trigonometry can be checked through the Vedic sutra, and that too without using pen and paper. Vedic maths systems are much easier and are time and space savers when compared with the conventional method. The Vedic mathematical formula might appear difficult at first sight but with practise, you will be attracted towards its beauty and simplicity. In a talk in 1958 at the Institute of Technology, Pasadena, California, Sri Bharti Krishna Tirthaji said, 'People who have practical knowledge of the application of the sutras, need not go in for the theory side of it at all.'

## To whom is Vedic mathematics beneficial?

Vedic mathematics is an emerging tool for students appearing in various competitive examinations like CAT, MAT, XAT, SAT, engineering, banking examinations where speed and accuracy play a vital role. Even students of IIT are said to be using this ancient technique for quick calculation. It is being taught in some of the most prestigious institutions in England and America. NASA scientists are said to have applied its principles in the area of artificial intelligence. Students, competitive exam aspirants, engineers, professionals, teachers and businessman are all reaping its benefits. Vedic mathematics is helpful to software developers in coding and programming too.

## What is the use of Vedic Mathematics when calculators and computers are available?

Vedic mathematics can speed up mathematical calculations such as arithmetic, algebra, trigonometry and geometry. Calculations are carried out mentally and students have the freedom to choose the best method for any particular situation. It reduces the time to solve a mathematical problem and boosts the student's confidence. In the present cut-throat competitive era, Vedic mathematics plays an important role in performing fast and effective calculations.

While calculating with a calculator, our involvement in the process in nil. This is highly dangerous in the long run, as we lose our ability to calculate in our daily life and our ability to think is also affected. Vedic mathematics emphasises on mental calculations and this keeps our brain fit and sensitive. Medical research has proved that our brain weight may increase by five percent if we do not have mental exercise. And brain weight once increased, cannot be reduced. According to a study conducted some years ago by a university in America, the constant use of calculators for more than twenty years by a person atrophies his brain significantly. Through Vedic mathematics we use both parts of our brain, thereby, keeping us mentally fit.

# Vedic Sutras

- Ekadhikena Purvena (एकाधिकेन पूर्वेण) – By one more than the previous one.

- Nikhilam Navatascaramam Dasatah (निखिलम् नवतश्चरमं दशत:) – All from nine and last from ten.

- Urdhva Tiryagbhyam (उर्ध्वतिर्यभ्याम्) – Vertically and crosswise.

- Paravartya Yojayet (परावर्त्योजयेत) – Transpose and Apply.

- Sunyam Samyasamuccaye (शून्यं साम्य समुच्चये) – The summation is equal to zero.

- Anurupye Sunyamanyat (आनुरूप्ये शून्यमन्यत्) – If one is in ratio, other one is zero.

- Sankalana-Vyavakalanabhyam (संकलन–व्यवकलनाभ्याम्) – By addition and subtraction.

- Puranapuranabhyam (पूरणापूरणाभ्याम्) – By completion and non-completion.

- Calana-Kalanabhyam (चलनकलनाभ्याम्) – Sequential motion.

- Yavadunam (यावदूनम्) – The deficiency.

- Vyastisamastih (ब्यष्टि समष्टि) – Whole as one and one as whole.

- Sesanyankena Caramena (शेषाण्यंकेन चरमेण) – Remainder by last digit.

- Sopantyadvayamantyam (सोपान्त्य द्वयमन्त्यम्) – Ultimate and twice the penultimate.

- Ekanyunena Purvena (एकन्यूनेन पूर्वेण) – By one less than the previous one.

- Gunitasamuccayah (गुणित समुच्चयः) – The whole product is sameGunakasamuccayah (गुणक समुच्चयः) – Collectivity of multipliers.

# Vedic Sub-sutras

- Anurupyena (आनुरूप्येण) – Proportionately.

- Sisyate Sesasamjnah (शिष्यते शेषसंज्ञ) – Knowing remainder from remainder.

- Adyamadyenantya-mantyena (आद्यमाद्येनान्त्यमन्त्येन) – First by first and last by last.

- Kevalaih Saptakam Gunyat (केवलैः सप्तकं गुण्यात) – Only multiple of seven.

- Vestanam (वेष्टनम्) – Osculation.

- Yavadunam-Tavadunam (यावदूनम तावदूनम) – Whatever be the deficiency, lessen it further.

- Yavadunam Tavadunikrtya Varganaca Yojayet (यावदूनम तावदूनम कृत्य वर्ग च योजयेत्) – Whatever the extent of its deficiency, lessen it further to that extent and set up the square of deficiency.

- Antyayotdasakepi (अन्त्ययोर्दशकेऽपि) – When the sum of the last digits is ten.

- Antyayoreva (अन्त्ययोरेव) – Only the last term.

- Samuccayagunitah (सामुच्चय गुणितः) – Sum of the coefficients in the product.

- Lopanasthapanabhyam (लोपस्थापनाभ्याम्) – By Elimination and Retention.

- Vilokanam (विलोकनम्) – The product of the sum of coefficient.

- Gunitasamuccayah Samuccaya gunitah (गुणित समुच्चय: समुच्चयगुणित:) – The product of the sum of the coefficients in the factor is equal to the sum of the coefficients in the product.

- Dwandwayogah (द्वंद्व योग) – Duplex combination.

- Shuddah (शुद्ध:) – Dot.

- Dhwajankam (ध्वजांकं) – Flag digit.

# 1

# Addition

## Introduction

The conventional method of mathematics taught in our schools for a long time, lacks speed and there is a greater chance of mistakes. The main problem is in transferring the carry-over digit from one column to another. Generally, in simple addition, whenever we get the sum of the digit in a particular column exceeding in the multiples of ten, we write the unit digit in the answer column and carry-over the remaining digit to the next column. The Vedic method discussed here avoids carrying over the digit to the next column and thus saves time.

## Vedic Method for Addition:

पूरणा पूरणाभ्याम्
(Puranapuranabhyam)

संकलन व्यवकलनाभ्याम्
(Sankalan Vyavkalanabhyam)

एकाधिकेन पूर्वेण
(Ekadhikena Purvena)

## Meaning of Vedic Sutra:

1. पूरणा पूरणाभ्याम् (Puranapuranabhyam): Completion of a base number in the multiple of 10. This formula helps the reader to form a group of two or more numbers in such a way that their unit digits add up to the multiple of 10.

2. संकलन व्यवकलनाभ्याम् (Sankalan Vyavkalanabhyam): This formula consists of two Sanskrit words-संकलन (Addition) and व्यवकलन (Subtraction).This sutra is applicable when the pair of numbers doesn't form a base (multiple of 10).

3. एकाधिकेन पूर्वेण (Ekadhikena Purvena): This sutra literally means one more than the previous one.

## Addition by Vedic Sutra

### पूरणा पूरणाभ्याम् (Puranapuranabhyam)

This method is useful for large calculations. The Vedic sutra is based on completion or non-completion of a base number that is the multiple of 10. The calculation (large calculations in particular) is made easier by pairing numbers which complete a base (i.e. in multiples of 10). The pairing becomes easier if you remember the complement of the digits. Here is a list of pairs that complement each other.

### Complementary (परममित्र संख्या) numbers

Two numbers are said to complement one another if their sum is 10.

0 and 10 are complementary to each other as 0 + 10 = 10
1 and 9 are complementary to each other as 1 + 9 = 10
2 and 8 are complementary to each other as 2 + 8 = 10
3 and 7 are complementary to each other as 3 + 7 = 10
4 and 6 are complementary to each other as 4 + 6 = 10
5 is complementary to 5 as 5 + 5 = 10

| Number | 0 | 1 | 2 | 3 | 4 | 5 | 6 | 7 | 8 | 9 |
|---|---|---|---|---|---|---|---|---|---|---|
| Complement | 10 | 9 | 8 | 7 | 6 | 5 | 4 | 3 | 2 | 1 |

**Rule:**

- Spot those numbers which if paired would result in a rounded result. The above complement table will help you in making such pairs immediately.
- Rearrange the numbers and add as per the pairing of numbers.

Let us take a few examples to understand the modus operandi clearly.

**Example:** Add: $26 + 59 + 394 + 66 + 11 + 14 = ?$

**Solution:**

Step 1: Here, the complete observation shows us that $26 + 14$ are likely to yield a rounded result. The same would be the case if 59 and 11 were paired. Moreover, the pair 394 and 66 yield a rounded result.

$$26 + 59 + 394 + 66 + 11 + 14$$

Step 2: Rearrange the pair and add as per the pairing done above.

$$= (26 + 14) + (59 + 11) + (394 + 66)$$
$$= 40 + 70 + 460$$
$$= 40 + 70 + 460$$

$$= (40 + 460) + 70$$
$$= 500 + 70$$
$$= 570$$

**Example 2**: Add: 456 + 361 + 244 + 119 + 11 = ?

**Solution**:

Step 1: Here the unit digit of 456 and 244 makes a rounded result. The same complement rule applies for 361 and 119. Arrange these pairs.

= 456 + 361 + 244 + 119 + 11

= (456 + 244 ) + (361 + 119) + 11

= (400 + 56 + 200 + 44) + (300 + 61 + 100 + 19) + 11

        [Split into smaller parts for easier calculation]

= (600 + 100) + (400 + 80) +11

= 700 + 480 + 11

= 1191

**Example**: Add: 36 + 5 + 23 + 2 + 14

Solution: Here 36 and 14 makes a rounded pair just as 23, 2 and 5 altogether makes a rounded figure. Arrange these pairs and add.

36 + (5 + 23 + 2) + 14

= 50 + 30

= 80

## संकलन व्यवकलनाभ्याम् (Sankalan Vyavkalanabhyam)

This formula consists of two Sanskrit words – संकलन (Addition) and व्यवकलन (Subtraction). This sutra is applicable when the pair of numbers doesn't form a base (i.e.multiple of 10). In this method we reach rounded off figures which logically make our calculation easier. The purpose of this sutra is to visualise and split a number as a function of two or more numbers that make the entire calculation easy.

Let us take a few examples to understand the modus operandi of this method.

a) $24 = 20 + 4$
b) $39 = 40 - 1$
c) $543 = 550 - 7 = 500 + 40 + 3$
d) $793 = 700 + 90 + 3 = 800 - 7$

The above splitting is done by choice and you are the best judge to decide what splitting will work for you. The splitting may be breaking a number as the sum of two/more numbers or difference of two numbers.

Now let us take a few examples of addition.

**Example:** Add $74 + 69$

**Solution:** $74 + 69$
$$= 70 + 4 + 70 - 1$$
$$= (70 + 70) + (4 - 1)$$
$$= 140 + 3$$
$$= 143$$

**Explanation:** First decompose each number into small parts; keeping in mind that at least one of the decomposed pairs is a multiple of 10.

**Example:** Add $324 + 296 + 159 + 43$

**Solution:** $(300 + 20 + 4) + (300 - 4) + (150 + 9) + (50 - 7)$
$$= (300 + 300 + 150 + 50) + (20 + 4 - 4 + 9 - 7)$$
$$= 800 + 22$$
$$= 822$$

**Example:** Add 596 + 498 + 345 + 765

**Solution:** (600 − 4) + (500 − 2) + (350 − 5) + (750 + 15)
    = (600 + 500) + (350 + 750) + (15 − 4 − 2 − 5)
    = 1100 + 1100 + 4
    = 2204

एकाधिकेन पूर्वेण (**Ekadhikena Purvena**)

The whole procedure of adding can be summarized in the following steps:

  a) Add the unit digits column-wise.
  b) When the running total becomes greater than 10, put a dot or tick on that number.
  c) Move ahead with the excess of ten and add it to the next digit of the column.
  d) Lastly, count the number of dots or ticks and note it down to the number next to the unit place figure and add the two.

**Example:** 486 + 654 + 987 = ?

**Solution:**

|               |   | 4 | 8 | 6 |
|---------------|---|---|---|---|
|               |   | $\overline{6}$ | $\overline{5}$ | $\overline{4}$ |
|               | + | 9 | $\overline{8}$ | 7 |
| Running total |   | 9 | 1 | 7 |

**Explanation:**

**In the first column:** 6 + 4 = 10, so take away the excess 0 (10 − 10 = 0) and add it with the next digit 7 of the first column. i.e. 0 + 7 = 7. Write the final sum 7 in the remainder column. Put a bar over 4.

**In the second column:** 8 + 5 = 13 , so write the excess 3 (15- 10 = 5) in the remainder column. Put a dot over 5. The excess will be further added to 8, making it equal to 11. Hence mark on 8 and write 1 in the answer column.

**In the third column:** 4 + 6 = 10, so mark on 6 and move with the excess 0 (10 − 10) to be added to the next number of the remainder column. Write 0 + 9 = 9 in the total column. Now count the number of bars in each column and place it down to the number next to the unit place as done in the first example and add the two to get the final result.

| Running total |   | 9 | 1 | 7 |
|---------------|---|---|---|---|
| Dots          | 1 | 2 | 1 |   |
| Sum           | 2 | 1 | 2 | 7 |

**Example:** 6489 + 5642 + 3241 = ?

**Solution:**

$$
\begin{array}{rrrr}
6 & 4 & 8 & 9 \\
\overline{5} & \overline{6} & \overline{4} & \overline{2} \\
+3 & 2 & 4 & 1 \\
\hline
4 & 2 & 6 & 2 \\
\end{array}
$$

**Explanation:**

**In the first column:** 9 + 2 = 11, so place a dot over 2 and add the excess 1 (11-10 = 1) to the next figure of the first column. i. e. 1 + 1 = 2.

**In the second column:** 8 + 4 = 12, so place a dot over 4 and add the excess 2 (12 − 10 = 2) to the next digit of the second column. i. e. 2 + 4 = 6

**In the third column:** 4 + 6 = 10, so place a dot over 6 and add the excess 0 (10 − 10) to the next digit of the column. i. e. 0 + 2 = 2

**In the fourth column:** 6 + 5 = 11, so place a dot over 5 and add the excess 1 (11 − 10 = 1) to the next digit of the column. i. e 1 + 3 = 4

Now count the number of dots in each column and place it down to the number next to the unit place as done in the first example and add the two to get the final result.

| Running total | | 4 | 2 | 6 | 2 |
|---|---|---|---|---|---|
| Dots | 1 | 1 | 1 | 1 | |
| Sum | 1 | 5 | 3 | 7 | 2 |

# 2

# Subtraction

## Introduction

The general meaning of subtraction is to remove certain things from a certain group. In other words, it is the process of finding a quantity which when added to one of the two given quantities will give the other. These quantities are respectively known as the subtrahend and minuend. The final result is called the remainder.

**Example:**

$$
\begin{array}{rl}
9\ 2\ 7\ 8 & \text{Minuend} \\
-\ 3\ 0\ 4\ 1 & \text{Subtrahend} \\
\hline
6\ 2\ 3\ 7 & \text{Remainder}
\end{array}
$$

Subtraction is considered a difficult mathematical operation and the subtraction operation would be slow and tiring with a carry-over at almost every step. Let us take a simple example to understand the process.

**Example 1:** Subtract 4768 from 8436

**Solution:**

$$
\begin{array}{r}
\overset{7}{}\ \overset{13}{}\ \phantom{0} \\
\overset{\cancel{8}}{}\ \overset{12}{}\ \phantom{0} \\
\overset{\cancel{2}}{}\ \overset{16}{} \\
\cancel{8}\ \cancel{4}\ \cancel{3}\ \cancel{6} \\
-\ 4\ 7\ 6\ 8 \\
\hline
3\ 6\ 6\ 8
\end{array}
$$

The general look of a copy will be the same. The number of steps involved in this operation obviously makes it a time consuming process. The operation can be simplified by either of the following techniques:

- Decompose both the numbers for easy calculation.
- Convert the subtraction to addition of a number.
- Carry out a digit-by-digit subtraction, without any carrying-over.

## Decomposition Method:

**Example 2**: Subtract 365 from 632

**Solution**:

$$
\begin{array}{rcl}
632 & = & 600 + 30 + 2 = 500 + 120 + 12 \\
-365 & = & 300 + 60 + 5 = -(300 + 60 + 5) \\
& = & 200 + 60 + 7
\end{array}
$$

This method is simple for a kindergarten student, but can't be used in the long run. The traditional method of borrowing as shown in Example 1 is tedious and cumbersome. Let us understand what method Vedic mathematics offers us to tackle this problem.

## Vedic Method for Subtraction:

| Vidic Method for Subtraction | |
| --- | --- |
| निखिलं नवतश्चरमं दशतः (Nikhilam Navatascaraman Dasatah) | मिश्रांक (Vinculum) |

Here we shall discuss the two Vedic methods that will be of immense use in solving subtraction in a few seconds and saving a lot of our time. Besides that, I shall lay emphasis on a special method that I have named the **Digit Separator Method**.

Meaning of Vedic Sutras:

1. Nikhilam Navatascaramam Dasatah (निखिलं नवतश्चरमं दशतः): The literal meaning of this Vedic Sutra is – **All from nine and last from ten.**
2. Vinculum (मिश्रांक) : In order to make the operation very simple and fast, we convert digits in a number,which are greater than 5, to the digits less than 5. After the conversion, the arithmetical operation becomes child's play.

Let us understand this further with a few examples—

## Nikhilam Navatascaramam Dasatah (निखिलं नवतश्चरमं दशतः)

This method works faster when subtraction is done from multiples of 10. The examples will give a clear understanding of the modus operandi to the readers. Let us first look at the given examples–

**Example:**

| 10000 | 40000 | 5900000 |
|---|---|---|
| −482 | −1172 | −48965 |

While calculating the above, several carry-overs are needed, which wastes time and confusion about accuracy still remains, The Vedic method helps you in this regard and saves your precious time.

**Rule:**

- Start moving from right to left. Replace every zero from the left with a 9 and the last zero with a 10. The extreme left digit before zero will get reduced by 1. Now do the simple subtracting without worrying about mistakes.

**Example:** Subtract 10000 – 462

**Solution:**

|       |        |          |
|-------|--------|----------|
| 10000 | will   | 9 9 9 10 |
| – 462 | become | – 4 6 2  |
|       |        | 9 5 3 8  |

**Example:** Subtract 40000 – 1172

**Solution:**

|        |        |           |
|--------|--------|-----------|
| 40000  | will   | 3 9 9 9 10 |
| – 1172 | become | – 1 1 7 2 |
|        |        | 3 8 8 2 8 |

Here, the extreme left digit i.e. 4, will get diminished by 1, and all the zeros thereafter will change into 9, except the last one. The last zero on the extreme right will be changed to 10.

Before we take a few more examples based on the above sutra, let me introduce to you the concept of the Complementary (परममित्र संख्या) Method.

Two numbers are said to be complement to one another if their sum is 10.

0 and 10 are complementary to each other as 0 + 10 = 10
1 and 9 are complementary to each other as 1 + 9 = 10
2 and 8 are complementary to each other as 2 + 8 = 10
3 and 7 are complementary to each other as 3 + 7 = 10
4 and 6 are complementary to each other as 4 + 6 = 10
5 is complementary to 5 as 5 + 5 = 10

| Number     | 0  | 1 | 2 | 3 | 4 | 5 | 6 | 7 | 8 | 9 |
|------------|----|---|---|---|---|---|---|---|---|---|
| Complement | 10 | 9 | 8 | 7 | 6 | 5 | 4 | 3 | 2 | 1 |

**Rule:**

- When the digit at minuend (upper digit) > subtrahend digit (lower digit), normal subtraction is done.
- In case the upper digit < lower digit, we take the complement of the difference as shown in the complement

table. The complement of the last digit is taken from 10 and the complements of the rest of the digits are taken from 9.

- When you arrive at a stage where there is no need to take the complement, subtract 1 extra from that column.

Let us take an example to understand this basic concept more clearly.

**Example:** 854 − 569 = ?

**Solution:**

$$
\begin{array}{r}
\text{From } 10 \\
8\ \ 5\ \ 4\ \bigg| \\
-\ \ 5\ \ 6\ \ 9\ \blacktriangledown \\
\hline
5
\end{array}
$$

Step 1: Since 4< 9, we take the complement of the difference of the digits. This complement will be taken from 10. The difference of 9 and 4 is 5. From the complement table it is evident that the complement of 5 is 5, so write 5 at the unit place.

Step 2: Again, 5 < 6, so we take the complement of the difference of two digits, but this time the complement is taken from 9 instead of 10. As the sutra suggests **all from 9 and last from 10**. The difference of 6 − 5 = 1 and its complement from 9 is 9 − 1 = 8. So write 8 at the ten's place.

$$
\begin{array}{r}
\text{From } 9 \\
8\ \ 5\ \bigg|\ 4 \\
-\ \ 5\ \ 6\ \blacktriangledown\ 9 \\
\hline
8\ \ 5
\end{array}
$$

Step 3: The difference of the digits at the hundred's place can easily be carried out as 8 > 5. So we don't need to take the complement for the third column. As we are now out of the complement, subtract 1 more in this column. Hence, instead of subtracting 8 − 5, we subtract 8 − 5 − 1 = 2.

$$
\begin{array}{r}
8\ 5\ 4 \\
-\ 5\ 6\ 9 \\
\hline
2\ 8\ 5
\end{array}
$$

**Example:** 8745 − 4599 = ?

**Solution:**

From 10
$$
\begin{array}{r}
8\ 7\ 4\ 5\ \big| \\
-\ 4\ 5\ 9\ 9\ \blacktriangledown \\
\hline
6
\end{array}
$$

Step 1: Since, 5 < 9, take the complement of the difference. 9− 5 = 4. The complement of 4 is 6.

Step 2: Again 4 < 9, so take the complement of the difference. The complement this time will be taken from 9.

Difference = 9 − 4 = 5 and complement of 5 from 9 is 9 − 5 = 4.

From 9
$$
\begin{array}{r}
8\ 7\ 4\ \big|\ 5 \\
-\ 4\ 5\ 9\ \blacktriangledown\ 9 \\
\hline
4\ \ \ \ 6
\end{array}
$$

Step 3: Now, 7 − 5 is easy to carry out and since we applied the complement in the next digit so  subtract 1 more in this column. Hence, the digit at the hundred's place will be 7 − 5−1 = 1

$$
\begin{array}{r}
8\ \ 7\ \ 4\ \ 5 \\
-\ 4\ \ 5\ \ 9\ \ 9 \\
\hline
1\ \ 4\ \ 6
\end{array}
$$

Step 4: Here, 8 − 4 = 4

$$
\begin{array}{r}
8\ \ 7\ \ 4\ \ 5 \\
-\ 4\ \ 5\ \ 9\ \ 9 \\
\hline
4\ \ 1\ \ 4\ \ 6
\end{array}
$$

**Example:** 87652 − 40269 = ?

**Solution:**

```
                    From 10
        8  7  6  5  2  |
     −  4  0  2  6  9  ↓
     ─────────────────
                    3
```

Step 1: 2 < 9, so take the complement of the difference from 10. Unit digit = 10 − (9 − 2) = 3

Step 2: 5 < 6, so take the complement of the difference from 9. Ten's digit = 9 − (6 − 5) = 8

```
                 From 10
        8  7  6  5  |  2
     −  4  0  2  6  ↓  9
     ─────────────────
                 8     3
```

Step 3: In the third column, we are out of complement as 6 > 2, so we subtract 1 more in that column and now the digit at the hundred's place = 6 − 2 − 1 = 3

```
        8  7  6  5  2
     −  4  0  2  6  9
     ─────────────────
              3  8  3
```

Step 4: Since 7 > 0, write 7 − 0 = 7 in the fourth column. Moreover, 8 > 4, so write 8 − 4 = 4 in the fifth column.

```
        8  7  6  5  2
     −  4  0  2  6  9
     ─────────────────
        4  7  3  8  3
```

**Example:** 459876 − 389924 = ?

**Solution:**

```
        4  5  9  8  7  6
     −  3  8  9  9  2  4
     ────────────────────
                    5  2
```

Step 1: 6 > 4 and 7 > 2, so normal subtraction will take place here. Hence,

Unit digit = 6 – 4 = 2 and

Ten's digit = 7 – 2 = 5

Step 2: Here, 8 < 9, so the Vedic sutra "All from 9 and last from 10" will be applied here. Now take the complement of the difference of the two numbers. The complement will be taken from 10.

The digit at the hundred place = 10 – (9 – 8) = 9

$$
\begin{array}{r}
\text{From 10} \\
4\ 5\ 9\ 8\ \big|\ 7\ 6 \\
-\ 3\ 8\ 9\ 9\ \big\downarrow\ 2\ 4 \\
\hline
9\qquad 5\ 2
\end{array}
$$

Step 3: Here, in the fourth column, 9 – 9 = 0; subtract the difference from 9.

The digit at the thousand's place = 9 – (9 – 9) = 9

$$
\begin{array}{r}
4\ \ 5\ \ 9\ \ 8\ \ 7\ \ 6 \\
-\ \ 3\ \ 8\ \ 9\ \ 9\ \ 2\ \ 4 \\
\hline
9\ \ 9\ \ 5\ \ 2
\end{array}
$$

Step 4: Again, 5 < 8, so subtract the difference from 9.

The digit at the ten thousand's place = 9 – (8 – 5) = 6

$$
\begin{array}{r}
4\ \ 5\ \ 9\ \ 8\ \ 7\ \ 6 \\
-\ \ 3\ \ 8\ \ 9\ \ 9\ \ 2\ \ 4 \\
\hline
6\ \ 9\ \ 9\ \ 5\ \ 2
\end{array}
$$

Step 5: In the sixth column, 4 > 3, which is easy to subtract but we had used the complement in the very next column so subtract one more in the sixth column.

i.e. 4 – 3 – 1 = 0

```
    4  5  9  8  7  6
 -  3  8  9  9  2  4
    _____
       6  9  9  5  2
```

मिश्रांक विधि (Vinculum method)

In order to convert a digit into a mishrank digit, the following steps are used:

- Subtract the given digit from 10 and place a bar over it.
- Apply Ekanyunena Purvena and add 1 to the digit on the left.

**Example:** 1. 438, on conversion to mishrank, will become 4 4 $\bar{2}$.

2. 4213883, on conversion, will become 4214$\overline{123}$.

**Explanation:** In the first example, the unit digit is more than 5 so in order to change 438 into mishrank, subtract 8 from 10 and put a bar on 2. Add 1 to the previous digit, 3, to get 4.

In the second example, subtract 8 in the ten's place from 10 and put a bar over 2. Add 1 to the previous digit 8, making it 9. Since it is again greater than 5, subtract 9 from 10 and put a bar over 1. Finally, the previous digit, 3, will become 4 on adding 1 to it.

Before I proceed with the examples on subtraction, let me first tell you how to convert a mishrank digit back to the original value.

- Subtract the mishrank digit marked with a bar from 10.
- Subtract 1 from the non-mishrank digit to the immediate left of the mishrank digit.

**Example:** 1. 44$\bar{2}$ will become 438

2. 4214$\overline{123}$ will become 4213883

**Explanation:** In the first example, the mishrank digit 2 will be subtracted from 10 (10 − 2 = 8)and its previous digit, 4, will get reduced by 1.

In the second example, the mishrank digit 2 will get subtracted from 10, making the digit at the ten's place = 10 − 2 = 8, and the mishrank digit to its immediate left, 1, will get subtracted from 9 (All from 9 and last from 10), making it 9 − 1 = 8. Now the digit at the thousand's place will get reduced by 1, i.e. 4 − 1 = 3.

## Subtraction by Mishrank Method

**Example:** Subtract 389 from 567

**Solution:** Place a bar over each digit on the subtrahend to signify that it is a negative number.

$$
\begin{array}{rccc}
 & 5 & 6 & 7 \\
+ & \bar{3} & \bar{8} & \bar{9} \\
\hline
 & 2 & \bar{2} & \bar{2} \\
\end{array}
$$

Here, $\quad 7 + \bar{9} = 7 - 9 = \bar{2}$

$\qquad 6 + \bar{8} = 6 - 8 = \bar{2}$

$\qquad 5 + \bar{3} = 5 - 3 = \bar{2}$

Now, convert $2\ \bar{2}\ \bar{2} = 1\ 7\ 8$

Apply, निखिलं नवतश्चरमं दशतः **(All from 9 and last from 10).** Here the mishrank digit 2 at the unit place will get subtracted from 10 i.e. 10 − 2 = 8 and the $\bar{2}$ on the immediate left will be subtracted from 9 i.e. 9 − 2 = 7. The digit at the hundred's place will get reduced by 1 making the 2 at hundred's place 1.

**Example:** Subtract 3699 from 4589

**Solution:** Place a bar over each digit on the subtrahend to signify that it is a negative number.

$$\begin{array}{rcccc} & 4 & 5 & 8 & 9 \\ + & \bar{3} & \bar{6} & \bar{9} & \bar{9} \\ \hline & 1 & \bar{1} & \bar{1} & 0 \end{array}$$

Here  $9 + \bar{9} = 0$

$8 + \bar{9} = \bar{1}$

$5 + \bar{6} = \bar{1}$

$4 + \bar{3} = 1$

Now, convert the mishrank digit back to original digit. Hence
$1\ \bar{1}\ \bar{1}\ 0 = 890$

Here, the mishrank digit 1 at the ten's place will be subtracted from 10, making it $10 - 1 = 9$, and the immediate left mishrank digit, 1, will be subtracted from 9, i.e. $9 - 1 = 8$. The digit at the thousand's place will get reduced by 1. i. e. $= 1 - 1 = 0$.

**Example:** Subtract $4568568 - 3478989$

**Solution:** Place a bar over each digit on the subtrahend to signify that it is a negative number.

|  |  | 4 | 5 | 6 | 8 | 5 | 6 | 8 |
|---|---|---|---|---|---|---|---|---|
|  | + | $\bar{3}$ | $\bar{4}$ | $\bar{7}$ | $\bar{8}$ | $\bar{9}$ | $\bar{8}$ | $\bar{9}$ |
| Mishrank digit |  | 1 | 1 | $\bar{1}$ | 0 | $\bar{4}$ | $\bar{2}$ | $\bar{1}$ |
| Original digit |  | 1 | 0 | 8 | 9 | 5 | 7 | 9 |

Here, the **All from 9 and last from 10** formulas is used to convert the mishrank digit to a non mishrank digit.
In the unit place, $\bar{1} = 10 - 1 = 9$
Ten's place $\bar{2} = 9 - 2 = 7$
Hundred's place $4 = 9 - 4 = 5$
Thousand's place $0 = 9 - 0 = 9$
Ten thousand's place $\bar{1} = 9 - 1 = 8$
(Since at this stage we are out of complement; 1 is subtracted from immediate left 1.)

## Digit Separator Method

The digit separator method is effective and works better for all types of subtraction problems. Before I take a few examples, let me introduce the modus operandi of this method.

**Rule:**

- Draw as many vertical separator lines as the number of digits
- Whenever the minuend at the top is less than the subtrahend at the bottom, put (−1) in the remainder column of each digit separator.
- Subtract algebraically, putting the sign of the larger number in the remainder column.
- Write the complement of the negative remainder.
- Club the digit and its next encircled digit from left to right to reach the final result.

**Example:** Subtract 254 − 168

**Solution:**

- Draw as many vertical digit separators

| 1$^{st}$ | 2$^{nd}$ | 3$^{rd}$ |
|---|---|---|
| 2 | 5 | 4 |
| −1 | 6 | 8 |

- Whenever the minuend at the top is less than the subtrahend at the bottom, put (−1) in the remainder column of each digit separator

| 1$^{st}$ | 2$^{nd}$ | 3$^{rd}$ | |
|---|---|---|---|
| 2 | 5 | 4 | -------- Minuend |
| −1 | 6 | 8 | -------- Subtrahend |
| | -1 | -1 | -------- Remainder |

Here $4 < 8$ and $5 < 6$, so the (-1) encircled is placed in $2^{nd}$ and $3^{rd}$ column.

- Subtract algebraically, putting the sign of larger number in the remainder column.

| $1^{st}$ | $2^{nd}$ | $3^{rd}$ | |
|---|---|---|---|
| 2 | 5 | 4 | -------- Minuend |
| -1 | 6 | 8 | -------- Subtrahend |
| 1 | ⊝-1 | ⊝-4 | -------- Remainder |

- Write the complement of the negative remainder in the $2^{nd}$ and $3^{rd}$ column. Here the complement of -1 and -4 is 9 and 6.

| $1^{st}$ | $2^{nd}$ | $3^{rd}$ | |
|---|---|---|---|
| 2 | 5 | 4 | -------- Minuend |
| -1 | 6 | 8 | -------- Subtrahend |
| 1 | ⊝9 | ⊝6 | -------- Remainder |

- Now subtract from left to right in the direction of the arrow.

1  ⊝  9  ⊝  6

$$= 086$$

**Example:** Subtract $5428 - 3765$

**Solution:**

- Draw as many vertical digit separator lines as the number of digits.

| $1^{st}$ | $2^{nd}$ | $3^{rd}$ | $4^{rd}$ |
|---|---|---|---|
| 5 | 4 | 2 | 8 |
| - 3 | 7 | 6 | 5 |
| | | | |

- Whenever the minuend at the top is less than the subtrahend at the bottom, put (–1) in the remainder column of each digit separator.

| 1$^{st}$ | 2$^{nd}$ | 3$^{rd}$ | 4$^{rd}$ | |
|---|---|---|---|---|
| 5 | 4 | 2 | 8 | --------- Minuend |
| – 3 | 7 | 6 | 5 | --------- Subtrahend |
| 2 | (–1) | (–1) | 3 | --------- Remainder |

Here, 2 < 6 and 4 < 7 in the 3$^{rd}$ and 2$^{nd}$ columns, which means that the minuend < subtrahend. Hence –1 encircled is placed in the respective remainder column.

- Now subtract as you do in algebra. i.e. put the sign of the larger number in the remainder column e.g. 4 – 7 = –3 in the 2$^{nd}$ column and 2 – 6 = –4 in the 3$^{rd}$ column.

| 1$^{st}$ | 2$^{nd}$ | 3$^{rd}$ | 4$^{rd}$ |
|---|---|---|---|
| 5 | 4 | 2 | 8 |
| – 3 | 7 | 6 | 5 |
| 2 | (–1) –3 | (–1) –4 | 3 |

- Write the complement of the negative remainder written in the 2$^{nd}$ and 3$^{rd}$ column. Here the complement of –3 and –4 are respectively 7 and 6.

| 1$^{st}$ | 2$^{nd}$ | 3$^{rd}$ | 4$^{rd}$ |
|---|---|---|---|
| 5 | 4 | 2 | 8 |
| – 3 | 7 | 6 | 5 |
| 2 | (–1) 7 | (–1) 6 | 3 |

- Now subtract from right to left as in the direction of the arrow.

$$2 \overset{\curvearrowright}{\enclose{circle}{-1}} \ 7 \ \overset{\curvearrowright}{\enclose{circle}{-1}} \ 3 \ \text{----Remainder Column}$$

= 1 6 6 3.

**Example:** Subtract 3 7 5 9 8 8 from 8 2 3 4 7 9.

**Solution:**

- Draw a digit separator vertical line.

| 1st | 2nd | 3rd | 4th | 5th | 6th | |
|-----|-----|-----|-----|-----|-----|---|
| 8 | 2 | 3 | 4 | 7 | 9 | --- Minuend |
| -3 | 7 | 5 | 9 | 8 | 8 | --- Subtrahend |
| | | | | | | |

- Subtract algebraically and write the remainder in the remainder column.

| 1st | 2nd | 3rd | 4th | 5th | 6th | |
|-----|-----|-----|-----|-----|-----|---|
| 8 | 2 | 3 | 4 | 7 | 9 | --- Minuend |
| -3 | 7 | 5 | 9 | 8 | 8 | --- Subtrahend |
| 5 | -5 | -2 | -5 | -1 | 1 | --- Remainder |

- In each of the remainder columns with negative sign, write -1 encircled. Here the 2nd, 3rd, 4th and 5th column has negative remainder.

| 1st | 2nd | 3rd | 4th | 5th | 6th | |
|-----|-----|-----|-----|-----|-----|---|
| 8 | 2 | 3 | 4 | 7 | 9 | ----- Minuend |
| -3 | 7 | 5 | 9 | 8 | 8 | ------ Subtrahend |
| 5 | $\enclose{circle}{-1}$-5 | $\enclose{circle}{-1}$-2 | $\enclose{circle}{-1}$-5 | $\enclose{circle}{-1}$-1 | 1 | ------ Remainder |

- Write the complement of each digit with negative sign. Here the complement of -5, -2, -5, and -1 in the 2$^{nd}$, 3$^{rd}$, 4$^{th}$ and 5$^{th}$ column are respectively 5, 8, 5 and 9.

| 1$^{st}$ | 2$^{nd}$ | 3$^{rd}$ | 4$^{th}$ | 5$^{th}$ | 6$^{th}$ | |
|---|---|---|---|---|---|---|
| 8 | 2 | 3 | 4 | 7 | 9 ----- Minuend | |
| −3 | 7 | 5 | 9 | 8 | 8 ------ Subtrahend | |
| 5 | −1 / 5 | −1 / 8 | −1 / 5 | −1 / 9 | 1 ------ Remainder | |

- Subtract from left to right in the direction of the arrows to get the required result. For the convenience of the reader, the remainder column is written here with an arrow sign.

$$= 5 \ (-1) \ 5 \ (-1) \ 8 \ (-1) \ 5 \ (-1) \ 9 \ 1$$
$$= 5\text{-}1 \ / \ 5\text{-}1 \ / \ 8\text{-}1 \ / \ 5\text{-}1 \ / \ 9 \ / \ 1$$
$$= 4 \ 4 \ 7 \ 4 \ 9 \ 1$$

# 3

# Multiplication

## Introduction

Multiplication is repeated addition. When we say 3 multiplied by 4, we simply say 3 + 3 + 3 + 3. The century-old method practiced all over the world has the unique pattern of multiplication that involves:                Multiplicand x Multiplier = Product

```
        4 2 3 5 ————————➤  Multiplicand
        X   2 5 ————————➤  Multiplier
      2 1 1 7 5
    + 8 4 7 0 X
    1 0 5 8 7 5 ————————➤ Product
```

This method is good, but there is no room for experimentation. We are bound to do the same in all types of multiplication. The process involves multiple stages and is error prone. A simple mistake is sufficient to make the calculation faulty. This is not all; a multiplication of 5 x 5 digits is bound to take at least 3-4 minutes. The best advantage of the Vedic Method over the present day calculation is that Vedic mathematics has ample fruit in its basket. You have different choices and you can choose the best possible method in the best situation. There are many special sutras that help you to find the answer of a special type of multiplication even in seconds and the *Urdhva Tiryagbhyam* method helps you to encounter all types of multiplication. So friends, jump into the ocean of Vedic sutras and gather the pearls of your own choice

# Vedic Sutras for Multiplication

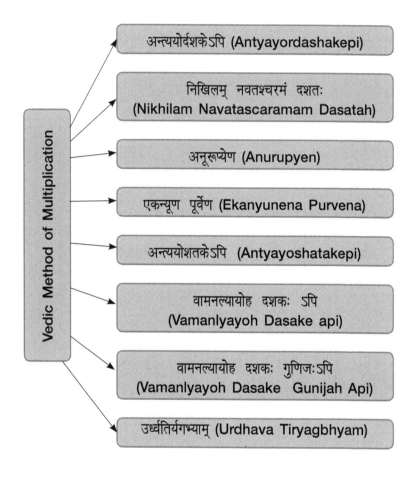

## Meaning of Vedic Sutra:

1. अन्त्ययोर्दशकेऽपि **(Antyayordashakepi):** The Vedic sutra is applicable when the sum of unit digit at multiplicand and multiplier is 10 and the remaining digits in the multiplicand as well as in the multiplier are the same.

2. निखिलम् नवतश्चरमं दशतः **(Nikhilam Navatascaramam**

**Dasatah)**: The literal meaning is – **All from 9 and last from ten**. This sutra works better when the numbers to be multiplied are very close to the base number. The base number should be the multiple of 10.

3. अनूरूप्येण **(Anurupyen)**: This Vedic sub-sutra literally means "**to proportionately**". This sub- sutra is applicable when either the multiplicand or multipliers is sufficiently very far from the power of 10.

4. एकन्यून पूर्वेण **(Ekanyunena Purvena)**: It literally means "**one less than the previous**". This sutra has limited applications. It is used for multiplication wherein the multiplier digits consist entirely of nines. It comes up under three types.

   a. When the multiplicand and multipliers each consist of equal number of digits
   b. When the multiplier has more digits than the multiplicand
   c. When the multiplicand has more digits than the multipliers

5. अन्त्ययोशतकेऽपि **(Antyayoshatakepi)**: The literal meaning is – **the sum of unit and ten's digit at multiplicand and multiplier is 100**. This sutra is valid as long as the sum of the last two digits is 100 and the remaining digits in multiplicand and multiplier are the same.

6. वामनल्यायोह दशकःऽपि **(Vamanlyayoh Dasake Api)**: This sutra is not taken from the original Vedic Text available, but is equally valuable as far as speedy calculation is concerned. The literal meaning is – **Sum of left two digits is equal to 10**. This sutra is valid if the digit at the unit place in both multiplicand and multiplier is equal and the sum of the left two digits is 10.

7. वामनल्यायोह दशकः गुणिजःऽपि **(Vamanlyayoh Dasake Gunijah Api)**: The literal meaning of this sutra is – **Sum of left two digits other than unit is a multiple of ten and**

**unit digits are the same.** This is also not a Vedic Sutra and is not taken from the Original Vedic Text.

8. उर्ध्वतिर्यग्भ्याम् (**Urdhava Tiryagbhyam**): It is a general formulae applicable to all cases of multiplication. It is a process of vertical and cross-wise multiplication. This method has been further simplified and dealt with Dot and Cross method in this book. A better understanding of this formula will also help you in multiplying two numbers with the other formulae mentioned above.

## Vedic Multiplication

### अन्त्ययोर्दशकेऽपि (Antyayordashakepi)

**Rule:**

- This formula has limited application and is valid as long as the sum of the unit digit at multiplicand and multiplier is 10 and the remaining digits are the same. The final product will consist of two parts – in the right hand part, the product of unit digit is to be written and in the left hand part, the product of remaining digits is to be written.

- Multiply the unit digits and write it in the RHS (Right Hands Side) .

- In the LHS (Left Hand Side), write the product of (Remaining Digit at Ten's/Hundred place) x (Remaining digit at Ten's /Hundred place + 1)

**Example** 1: Multiply 24 by 26

**Solution:**

$$\left. \begin{array}{r} 2\ 4 \\ \underline{x\ 2\ 6} \end{array} \right\} \text{Here } 4 + 6 = 10$$

Here the sum of the unit digit is 4 + 6 = 10 and the ten's digit in multiplicand and multiplier are the same. Put 4 x 6 = 24 in RHS and 2 x (2+1) = 6 in LHS

Hence, 24 x 26 = 624

**Example 2:** Multiply 62 by 68

**Solution:** Sum of unit digit = 2 + 8 = 10
Ten's digit in Multiplicand and Multiplier = 2
Hence,
LHS = 2 x 8 = 16
RHS = 6 x (6 + 1) = 42
Result = 4216

**Example 3:** Multiply 93 by 97

**Solution:** Sum of unit digit = 3 +7 = 10
Ten's digit in Multiplicand and Multiplier = 9
Hence,
LHS = 3 x 7 = 21
RHS = 9 x (9 + 1) = 90
Result = 9021

## निखिलम् नवतश्चरमं दशतः (Nikhilam Navatascaramam Dasatah)

This sutra works better when both the multiplicand and multiplier are very close to the base. The base should be in the form of $10^n$, where n is a natural number.

**Rule:**

- Write the two numbers to be multiplied above and below at the right side of your notebook.
- Write the deviation of multiplicand and multiplier from the base and place them next to the digit to be multiplied.
- The final result will have two parts.

a) The left hand part will be obtained by cross operation of two numbers written diagonally.

b) The right side of the answer will be obtained by multiplying the deviations.

- The number of digits in the right hand part will be in accordance to the number of zeros in the base number. In simple words, if the base is 100, the right hand part will have two digits and if the base is 1000, the right hand part will have three digits.

- In case there is lesser number of digits in the right side, accommodate as many zeros before the right hand part so that the total number of digits in that part is equal to the number of zeros in the base.

- Here is the table that will guide you in deciding the number of digits to be placed on the right hand side.

| Base | Number of digits at the right side of vertical line. | |
|---|---|---|
| 10 | 1 | 0 |
| 100 | 2 | 00 |
| 1000 | 3 | 000 |
| 10000 | 4 | 0000 |
| 100000 | 5 | 00000 |
| 1000000 | 6 | 000000 |
| 10000000 | 7 | 0000000 |

Let us take a few examples to understand the modus operandi of the above Vedic Sutra.

*Case 1: When both numbers are below the base.*

**Example 4:** Multiply 8 by 7

**Solution:**

(a) Put the multiplicand and multiplier as shown here.

8
x 7

(b) Both the numbers are closer to the base 10, so take Base
    = 10.
    Deviation of 8 = 8 – 10 = – 2
    Deviation of 7 = 7 – 10 = – 3

(c) Put the deviation at the right side along with the number
    to be multiplied.
    8  – 2
    x 7  – 3

(d) Write the left hand digit by cross operation of any of
    two diagonal. Here 8 – 3 = 5 and 7 – 2 = 5
    8    – 2
    x  7  – 3
       5 /

(e) The right hand digit will be the multiplication of the
    deviation. The product of deviation is (–2) x (–3) = 6.
    8   – 2
    7   – 3
    5 /  6

**Example 5:** Multiply 95 by 91

**Solution:**

a) Put the multiplicand and multiplier as shown here.

   95
   x 91

b) Both the numbers are closer to the base 100, so take
   Base = 100.
   Deviation of 95 = 95 – 100 = – 5
   Deviation of 91 = 91 – 100 = – 9

c) Put the deviation at the right side along with the number
   to be multiplied.

```
    95  – 5
x  91  – 9
```

d) Write the left hand digit by cross operation of any of the two diagonals. Here 95 – 9 = 86 or 91 – 5 = 86 is written in the left hand part.

```
    95    – 5
x  91    – 9
86 /
```

e) The right hand digit will be the multiplication of the deviation.

```
    95  – 5
x  91  – 9
86 /45
```

*Case 2: When both the numbers are above the base.*

**Example 6:** Multiply 15 by 11

**Solution:**

a) Put the multiplicand and multiplier as shown here.

```
    15
x  11
```

b) Both the numbers are closer to the base 10, so take Base = 10.
   Deviation of 15 = 15 – 10 = 5
   Deviation of 11 = 11 – 10 = 1

c) Put the deviation at the right side along with the number to be multiplied.

```
    15  + 5
x  11  + 1
```

d) Write the left hand digit by cross operation of any of the two diagonals.

```
    15    +5
x  11    +1
16 /
```

e) The right hand digit will be the multiplication of the deviation.

$$15 \quad +5$$
$$\underline{\times 11 \quad +1}$$
$$16 \ / \ 5$$

**Example 7**: Multiply 105 by 104

**Solution**:

a) Put the multiplicand and multiplier as shown here.

$$105$$
$$\underline{\times 104}$$

b) Both the numbers are closer to the base 100, so take Base = 100.

Deviation of 105 = 105 − 100 = +5
Deviation of 104 = 104 − 100 = + 4

c) Put the deviation at the right side along with the number to be multiplied.

$$105 \quad + 5$$
$$\underline{\times 104 \quad + 4}$$

d) Write the left hand digit by cross operation of any of the two diagonals.

$$105 \quad + 5$$
$$\underline{\times 104 \quad + 4}$$
$$109 \ /$$

e) The right hand digit will be the multiplication of the deviation.

$$105 \quad + 5$$
$$\underline{104 \quad + 4}$$
$$109 \ / \ 20$$

*Case 3: When one number is above the base and another is less than the base.*

**Example 8**: Multiply 12 by 8

**Solution:**

a) Put the multiplicand and multiplier as shown here.

```
  12
x  8
```

b) Both the numbers are closer to the base 10, so take Base = 10.

Deviation of 12 = 12 – 10 = +2

Deviation of 8 = 8 – 10 = –2

c) Put the deviation at the right side along with the number to be multiplied.

```
  12  + 2
x  8  – 2
```

d) Write the left hand digit by cross operation of any of the two diagonals.

```
  12  + 2
x  8  – 2
10 /
```

e) The right hand digit will be the multiplication of the deviation. The product of (+2) x ( –2) = –4 is written in the RHS.

```
  12 + 2
x 8  – 2
10 / –4
```

e) When there is a minus (–) sign at the right hand product, use the Nilhilam formulae which states, "All from 9 and the last from 10." Hence subtract the right hand digit (–4) from 10 and left hand part will get diminished by 1. i.e. 10 – 1 = 9

```
   12 + 2
x   8 – 2
   10 / –4
= 9 / 10 – 4
= 9 / 6
```

**Example 9:** Multiply 122 by 98

**Solution:**

a)  Put the multiplicand and multiplier as shown here.

    122
    x  98

b)  Both the numbers are closer to the base 100, so take
    Base = 100.
    Deviation of 122 = 122 − 120 = +22
    Deviation of 98 = 98 − 100 = −2

c)  Put the deviation at the right side along with the number
    to be multiplied.

    122  + 22
    x 98  −  2

d)  Write the left hand digit by cross operation of any of the
    two diagonals.

    122  + 22
    X98  −  2
    120 /

e)  The right hand digit will be the multiplication of the
    deviation.

    122 + 22
    x  98 −  2
    120 / −44

f)  When there is a minus (−) sign at the right hand product,
    use the Nilhilam formulae which states, "All from 9 and
    the last from 10." Hence subtract the right hand digit
    (−44) from 100 and left hand part 120 will get diminished
    by 1. i.e. 120 − 1 =119

    122  + 22
    x  98  −  2
    119  / 100 − 44
    = 119  / 56

*Case 4: Adjustment of right side digit of the product.*

Two sub-cases may arise here:

a) When the number of digits on the right hand side is more than the permissible limit.

b) When the number of digits on the right hand side is less than the permissible limit.

*Sub case (a): When the number of digits on the right hand side is more than the permissible limit.*

**Example 10:** Multiply 16 by 15

**Solution:**

a) Put the multiplicand and multiplier as shown here.

    16
    x  15

b) Both the numbers are closer to the base 10, so take Base = 10.

Deviation of 16 = 16 – 10 = +6

Deviation of 15 = 15 – 10 = +5

c) Put the deviation at the right side along with the number to be multiplied.

    16  + 6
    x  15  + 5

d) Write the left hand digit by cross operation of any of the two diagonals.

    16   +6
    x 15   +5
    21 /

e) The right hand digit will be the multiplication of the deviation.

    16  + 6
    x 15  + 5
    21 / 30

Here, the number of digit in RHS is two, which is more than the permissible number of digits in RHS (See Table 1). The number of permissible digits in RHS should be in accordance with the base number. Since, the base is 10, the number placed at the right side should be of one digit. In such a case, we transfer the extreme left digit of RHS to the LHS and add them.

$$16 \ + \ 6$$
$$\underline{x \ 15 \ + \ 5}$$
$$21 \ / \ 30$$

= 240

**Example 11:** Multiply 13 by 18

**Solution:**

a) Put the multiplicand and multiplier as shown here.

13
x 18

b) Both the numbers are closer to the base 10, so take Base = 10.

Deviation of 13 = 13 – 10 = +3
Deviation of 18 = 18 – 10 = +8

c) Put the deviation at the right side along with the number to be multiplied.

13  +3
x 18  +8

d) Write the left hand digit by cross operation of any of the two diagonals.

13  +3
x 18  +8
21 /

e) The right hand digit will be the multiplication of the deviation.

$$13 + 3$$
$$18 + 8$$
$$21 / 24$$

Since, the base is 10, the number placed at the right side should be of one digit, so transfer the extreme left digit of RHS to the LHS and add them.

$$\begin{array}{r} 13 \ +3 \\ \times\ 18\ +8 \\ \hline 21\ /\ 24 \end{array}$$

$$= 234$$

*Sub case (b): When the number of digits on the right hand side is less than the permissible limit.*

**Example 12:** Multiply 96 by 98

**Solution:**

   a)  Put the multiplicand and multiplier as shown here.

       96
      x 98

   b)  Both the numbers are closer to the base 100, so take Base = 100.

      Deviation of 96 = 96 – 100 = – 4

      Deviation of 98 = 98 – 100 = – 2

   c)  Put the deviation at the right side along with the number to be multiplied.

       96  – 4
      x 98  – 2

   d)  Write the left hand digit by cross operation of any of the two diagonals.

       96  – 4
      x 98  – 2
       94 /

e) The right hand digit will be the multiplication of the deviation.

96 −4
98 −2
94 / 8

Since, the base is 100, the number placed at the right side should consist of two digits. But there is a single digit in the RHS. In such a case, we place the zero to the left in RHS so that the total number of digits in RHS is equal to the permissible number of digits. See Table 1 for better understanding.

96 −4
98 −2
94 / 08

**Example 13:** Multiply 989 by 995

**Solution:**

a) Put the multiplicand and multiplier as shown here.
   989
   x 995

b) Both the numbers are closer to the base 1000, so take Base = 1000.
   Deviation of 989 = 989 − 1000 = −11
   Deviation of 995 = 995 − 1000 = − 5

c) Put the deviation at the right side along with the number to be multiplied.
   989 −11
   x 995 −5

d) Write the left hand digit by cross operation of any of the two diagonals.
   989 −11
   x 995 −5
   984 /

e) The right hand digit will be the multiplication of the deviation.

$$
\begin{array}{r}
989 \quad -11 \\
\times\ 995 \quad -5 \\
\hline
984\ /\ 55
\end{array}
$$

Since the base is 1000, the number placed at the right side should consist of three digits. So in order to meet the requirement of permissible digit in RHS, we place the zero to the left in the RHS

(Refer Table 1).

$$
\begin{array}{r}
989 \quad -11 \\
995 \quad -5 \\
\hline
984\ /\ 055
\end{array}
$$

Till now, we have seen examples in which both the numbers were closer to the base. Now let us consider a case where the two numbers are nearer to a different base. Hey, are you worried? Don't panic, the problem will be solved in a similar fashion with a slight change in the LHS.

• Write the numbers with their respective deviations from the base as done earlier.
• Write the base of each number in a bracket and cancel an equal number of zeros in the bracket.
• The RHS will be calculated as done above by placing the product of deviations, and will have the number of digits equal to the number of zeros cancelled.
• In the LHS, write the sum of the cross product of the first diagonal and the deviation of second number.

**Example 14:** Multiply 107 by 1008

**Solution:** Here, the two numbers are of different base. 107 is closer to base 100 and 1008 is closer to base 1000. Hence, the respective deviations of the numbers are $+7$ ($107 - 100$) and $+8$ ($1008 - 1000$).

```
        Deviation        Base
   107   +7              (100)
 x 1008  +8            (1000)
```

- Cancel equal number of zeros of the different bases.

```
        Deviation        Base
   107   +7              (1ØØ)
 x 1008  +8            (1ØØØ)
```

- LHS = 107 x 10 + 8 = 1078

  RHS = 7 x 8 = 56

- Hence, 1008 x 107 = 107856

## Anurupyena Sutra (आनुरूप्येण)

The word Anurupyena simply means **"Proportionately."** This method is applicable only when the multiplicand and multipliers are very far from the theoretical base. In other words, when the multiplicand and multipliers are distant from the power of 10, we use this sutra.

Let us understand the modus operandi of the Anurupyena sutra. In order to avoid a tragic situation, we take two types of bases:– one is a theoretical base and other is a working base. The theoretical base is taken in the power of 10 and the working base is taken as the multiple of 10. Let us see this example to make the modus operandi clear.

*Sub-case 1: When the left hand figure is completely divisible by the divisor of the working base.*

**Example 15:**  Multiply 48 by 42

**Solution:**

Theoretical base = 100

Working base = 100 | 2 = 50

Deviation of 48 from the working base = 48 – 50 = – 2
Deviation of 42 from the working base = 42 – 50 = – 8

$$48 \quad – 2$$
$$\underline{\text{x } 42 \quad – 8}$$

The working procedure is almost the same except for the fact that the left hand side figure will be divided by 2 as our working base is half of the theoretical base. As far as the right hand figure is concerned, it will remain unaffected.

$$48 \quad\searrow\quad\; – 2$$
$$\underline{\text{x } 42 \quad\nearrow\quad – 8}$$
$$40 \mid 16$$
$$= \tfrac{1}{2} \text{ x } 40 \mid 16$$
$$= 2016$$

**Example 16**: Multiply 494 by 488

**Solution**:

Theoretical Base = 1000
Working Base = 500 = 1000/2
Deviation of 494 from the working base = 494 – 500 = – 6
Deviation of 42 from the working base = 488 – 500 = – 12

Do the operation as described in Nikhilam method, with a slight difference that the left hand figure will be divided by 2 as our working base is half of the theoretical base.

$$494 \quad\searrow\quad\; – 6$$
$$\underline{\text{x } 488 \quad\nearrow\quad – 12}$$
$$482 \mid 72$$

Since the theoretical base is 1000, the number of digits in the RHS will be 3, therefore put a zero behind 72.

$$494 \quad\searrow\quad\; – 6$$
$$\underline{\text{x } 488 \quad\nearrow\quad – 12}$$
$$482 \mid 072$$

Now, divide the LHS by 2

$$
\begin{array}{r}
494 \quad \diagdown \quad -6 \\
\underline{\text{x } 488 \quad \diagup \quad -12} \\
2 \ ) \ \overline{482 \ | \ 072} \\
= 241 \ / \ 072
\end{array}
$$

*Sub-case 2: When the left hand figure, on division by the divisor of working base gives a fractional quotient.*

**Example 17**: Multiply 48 by 49

**Solution**:

Theoretical base = 100
Working base = 50 = (100 | 2)

Step 1: Write the deviation (–2) and (–1) against the number 48 and 49 taken from its working base.

$$48 - 50 = -2 \qquad \text{and} \qquad 49 - 50 = -1.$$

$$
\begin{array}{r}
48 \ -2 \\
\underline{\text{x } 49 \ -1}
\end{array}
$$

Step 2: Do the required operation by taking any diagonal and writing the result at the left side of the vertical line. Multiply the right hand figure vertically.

$$
\begin{array}{r}
48 \quad \diagdown \quad -2 \\
\underline{\text{x } 49 \quad \diagup \quad -1} \\
47 \ | \ 02
\end{array}
$$

Step 3: Divide the left hand figure by 2. Here, 47, on dividing by 2, give us a fractional quotient. i.e. 23 ½ .

Step 4: The fractional part ½ (i.e. ½ of the theoretical base 100 = 50) is taken over to the right hand side.

$$48 \diagdown \quad -2$$
$$\underline{\times\ 49 \diagup \quad -1}$$
$$2\ )\ 47\ |\ 02$$
$$=23\ ½\ |\ 02$$
$$=23\ |\ 50+\ 2$$
$$=2352$$

**Example 18:** Multiply 251 by 252?

**Solution:**

Theoretical base = 1000
Working base = 250= (1000 | 4)

Step 1: Deviation of 251 = 251 – 250 = + 1
Deviation of 252 = 252 – 250 = + 2

Write the deviation + 2 and +1 against the numbers 251 and 252 taken from its working base.

$$251 \quad + 2$$
$$\underline{\times\ 252 \quad + 1}$$

Step 2: Take any diagonal and do the required operation. Write the result in LHS. Multiply the right hand figure vertically.

$$251 \diagdown \quad + 1$$
$$\underline{\times\ 252 \diagup \quad + 2}$$
$$253\ |\ 2$$

Step 3: Divide the left hand figure by 4. Here, 253, on dividing by 4, give us a fractional quotient. i.e. 63 ¼.

Step 4: The fractional part ¼ (i.e. ¼ of the theoretical base 1000 = 250) is taken over to the right hand side. Moreover, the number of digits in the RHS should be re-written according to the theoretical base. Since there are three zeros in the theoretical base,place two zeros before 2 in RHS so that the number of digits in RHS is equal to the number of zeros in the theoretical base.

$$251 \quad \nearrow \quad +1$$
$$\underline{\times\ 252} \quad \searrow \quad +2$$
$$4\ \overline{)\ 253\ |\ 002}$$
$$=\ 63\ \tfrac{1}{4}\ |\ 002$$
$$=\ 63\ |\ 250+002$$
$$=\ 63252$$

*Sub-case 3: When the right hand vertical product is negative.*

**Example 19**: Multiply 52 by 48

**Solution**: Theoretical base = 100

Working base = 50 = (100 | 2)

As discussed earlier, the excess or the deficiency from the working base is written against the number.

52 = 50 + 2 and 48 = 50 – 2

$$52 \quad +2$$
$$\underline{\times\ 48 \quad -\ 2}$$

Step 1: Perform the desired operation diagonally and vertically.

$$52 \quad \nearrow \quad +2$$
$$\underline{\times\ 48 \quad \searrow \quad -\ 2}$$
$$50\ |\ -04$$

Step 2: Divide the left hand figure by 2. Here 2 is the divisor of the theoretical base.

$$52 \quad +2$$
$$\underline{\times\ 48 \quad -2}$$
$$2\ \overline{)\ 50\ |\ -04}$$
$$=25\ |\ -04$$
$$24\ |\ 96$$

When the right hand figure is negative, the "**Nikhilam Navatas caramam dasatah**" formulae will be used. This simply directs us to subtract 1 from the left hand figure and to subtract the right hand figure from the theoretical base.

**Example 20**: Multiply 512 by 494

**Solution**: Theoretical base = 1000

Working base = 500 = (1000 | 2)

As discussed earlier, the excess or the deficiency from the working base is written against the number.

512 = 500 + 12  and  494 = 500 – 6

    512   + 12
  x 494   – 6

Step 1: Perform the desired operation diagonally and vertically.

    512          + 12
  x 494          – 6
    506  | –072

Step 2: Divide the left hand figure by 2

        512    + 12
      x 494    –6
  2 ) 506 |  – 072
    = 253 |  – 072
    = 252 | 1000 – 72
    = 252 | 928

This sutra works equally well when the base is the multiple of 10 (20, 30, 40, 60, 70...), multiple of 100 (200, 300, 400,) etc. The modus operandi of such multiplication is slightly different. Instead of dividing the LHS figure by 2, 4, 6, or 8, multiply the LHS by the sub-base number. If you are taking the base = 20 = 2 x 10; then base = 10 and sub- base = 2.

**Example 21**: Multiply 41 by 48?

**Solution**:

```
41        ⟋⟍ + 1
x 48     ⟋  ⟍ + 8   |⟋⟍        (Working base: 10 X 4 =40)
49 | 8
=49 | 8
x 4
=196| 8
```

**Example 22**:  Multiply 252 by 298

**Solution**:

```
252     ⟋⟍ - 4 8
x 299  ⟋  ⟍ - 0 1  |⟋⟍        (Working base: 100 X 3 =300)
251 | 48
 x 3
=753 | 48
=75348
```

**Example 23**: Multiply 687 by 695

**Solution**:

```
687    ⟋⟍ - 013
x 695 ⟋  ⟍ - 005  |⟋⟍        (Working base: 100 X 7 =700)
682 | 65
 x 7
=4774 | 65
=477465
```

**Example 24**: Multiply 889 by 895

**Solution**:

```
889    ⟋⟍- 011
x 895 ⟋  ⟍- 005  |⟋⟍        (Working Base = 900= 9 x 100)
884 | 055
x 9
 7956 | 055
= 7956055
```

## Ekanyuena Purvena

This one liner multiplication technique is a perfect beauty of Vedic Mathematics. A few months ago, I had gone to Ahmedabad to attend a Vedic Mathematics seminar. While addressing the gathering of students and teachers, I wrote a 9 digit number on the blackboard and asked the audience to multiply the written digit with 9 times 9.

$$569876943 \times 999999999 = ?$$

I had even allowed them to use the calculator, but the calculator showed an error message. I asked the audience to multiply the numbers manually and in the mean time, I wrote 569876942430123057 on the blackboard. The audience took more than 5 minutes and I had taken less than 15 seconds to write down the answer. The audience was amazed to see that the result I had written in less than 15 seconds was absolutely correct.

The Vedic Sutra – Ekanyuena Purvena is simply an awesome method of multiplying two numbers but this has limited application. This sutra work only under three conditions:

1) When the number of digits in the multiplicand and number of 9s in multipliers is the same.
2) When the number of 9s in the multipliers are more than the number of digits in the multiplicand.
3) When there is less number of 9s in the multiplier than the number of digits in the multiplicand.

Now let us take each case one by one.

### Case 1: *When the number of digits in the multiplicand and the number of 9s in the multipliers is the same.*

**Rule:**

• Subtract 1 from the multiplicand and write the result in LHS.

- Subtract the multiplicand by applying Nikhilam Navatascaramam Dasatah Vedic sutra and write the result in RHS.

**Example 25:** Multiply 6543 by 9999

**Solution:** Here, the number of digits in the multiplicand is equal to the number of 9s in the multipliers. As the rule suggests, the answer will have two parts.

LHS = Multiplicand − 1 = 6543 − 1 = 6542

RHS = Apply the Nikhilam method of subtraction and subtract the unit digit from 10 and the rest of the digits from 9. We get, 9 − 6 = 3, 9 − 5 = 4, 9 − 4 = 5 and, 10 − 3 = 7. Thus RHS will have 3457.

In order to simplify the calculation in RHS, we may subtract the result obtained in LHS from the multiplier.

RHS = 9999 − 6542 = 3457

Hence, 6543 x 9999 = 65423457

**Example 26:** Multiply 89654876 by 99999999

**Solution:**

LHS = 89654876 − 1 = 89654875

RHS = 99999999 − 89654875 = 10345124

Hence, 89654876 x 99999999 = 8965487510345124

**Example 27:** Multiply 83465087629 by 99999999999

**Solution:**

LHS = 83465087629 − 1 = 83465087628

RHS = 99999999999 − 83465087628 = 16534912371

Hence, 83465087629 x 99999999999 =

8346508762816534912371

**Example 28:** Multiply 45682 by 99999

**Solution:**

LHS = 45682 − 1 = 45681

RHS = 99999 – 45681 = 54318

Hence, 45682 x 99999 = 4568154318

With a little practice, you can write the RHS manually in no time. Subtract each digit of LHS from 9. If LHS = 23, then RHS will be 76.

**Case 2:** *When the number of 9s in the multipliers is more than the number of digits in the multiplicand.*

**Rule:** In case 2, the same procedure will be applied as in case 1.

Let us take a few examples.

**Example 29:** Multiply 456 by 9999

**Solution:**

> LHS = 456 – 1 = 455
> RHS = 9999 – 455 = 9544
> Hence, 456 x 9999 = 4559544

**Example 30:** Multiply 56892 by 9999999

**Solution:**

> LHS = 56892 – 1 = 56891
> RHS = 9999999 – 56891 = 9943108
> Hence, 56892 x 9999999 = 568919943108

**Example 31:** Multiply 13324 by 99999999

**Solution:**

> LHS = 13324 – 1 = 13323
> RHS = 99999999 – 13323 = 99986676
> Hence, 13324 x 99999999 = 1332399986676

**Case 3:** *When there is less number of 9s in the multiplier than the number of digits in the multiplicand.*

This case is a little bit different from the last two cases discussed so far under Ekanyena Purvena. In order to get the result you have to –

a) Add as many zero as the numbers of 9s to the multiplicand.

b) Subtract the original multiplicand from the figure obtained in Step 1.

**Example 32:** Multiply 1564 by 99

**Solution:**

The multiplicand 1564 has 4 digits, whereas there are two 9s in the multiplier.

```
  1 5 6 4
X 9 9
```

1) Since there are two 9s, put two zeros at the end of 1564, making it 156400.

2) Subtract (1564 original multiplicand) from 156400

i.e.
```
  1 5 6 4 0 0
 -1 5 6 4
  1 5 4 8 3 6
```

Hence, 1564 x 99 = 154836

**Example 33:** Multiply 783459 by 9999

**Solution:**

1) Since there are four 9s, put four zeros at the end of 783459, making it 7834590000

2) Subtract the original number 783459 from 7834590000.

```
  7 8 3 4 5 9 0 0 0 0
 -        7 8 3 4 5 9
  7 8 3 3 8 0 6 5 4 1
```

Hence, 783459 x 9999 = 7833806541

**Example 34:** Multiply 45678 by 999

**Solution:**

1) Since there are three 9s, put three zeros at the end of 45678, making it 45678000.

2) Subtract the original number from 45678000.

$$45678000$$
$$\underline{-\ 456778}$$
$$45632322$$

Hence, 45678 x 999 = 45632322

This can be done by the Eknyuenane Purvena method effortlessly after a little practice. Let's see how it works. In this method:

a) Subtract 1 from the original number and place it in LHS.

b) Write as many digits from right to left equal to the number of 9s that of multiplicand in RHS. Suppose you have to multiply 147 by 99 so RHS part of answer should contain 2 digits 47 taken from right, equal to the number of 9 in the multiplier.

c) The remaining digits in the original number, after removing the digits from the right to left, placed in the RHS should be subtracted in the LHS.

d) Write the complement of the digits placed in the RHS by applying the Nikhilam sutra.

**Example:** Multiply 147 by 99

**Solution:** Since there are two 9s in the multiplier, two digits from right to left of the multiplicand will be placed in the RHS. In the LHS, subtract 1 from the original number.

LHS = 147 − 1 = 146

RHS = Complement of 47

Now subtract the remaining digits i.e. 1 from LHS and write the complement of 47 in RHS.

LHS = 146 − 1 = 145

RHS = 100 − 47 = 53

Hence, 147 x 99 = 14553

**Example:** Multiply 259648 by 9999.

**Solution:** Since there are four 9s in the multiplier, four digits from right to left of multiplicand i.e. 9648 will be placed in the RHS. In LHS, subtract 1 from the original number.

LHS = 259648 - 1 = 259647
RHS = Complement of 9648

Now subtract the remaining digits i.e. 25 from LHS and write the complement of 9648 in RHS.

LHS = 259647 - 25 = 259622
RHS = 10000 - 9648 = 0352
Hence, 259648 x 9999 = 2596220352

**Example:** Multiply 52876 by 99

**Solution:** Since there are two 9s in the multiplier, two digits from right to left of the multiplicand i.e. 76 will be placed in the RHS. In LHS, subtract 1 from the original number.

LHS = 52876 - 1 = 52875
RHS = Complement of 76

Now subtract the remaining digits i.e. 528 from LHS and write the complement of 76 in RHS.

LHS = 52875 - 528 = 52347
RHS = 100 - 76 = 24
Hence, 52876 x 99 = 5234724

## अन्त्ययोशतकेऽपि (Antyayoshatakepi):

**Rule:**

- This sutra is applicable when the sum of the last two digits (unit and ten's) in the multiplicand and multiplier is 100 and the rest of the digits are the same.
- Multiply the last two digits and write the product in RHS
- Multiply the remaining digits that are the same, with the next digit i.e digit + 1 and write the product in the LHS. In simple word, if the digit at the hundred's place is 7, multiply it with the next digit 8 and write the product in LHS.

**Example 35:** Multiply 782 by 718

**Solution:** Here the sum of the last two digit of the multiplicand and multiplier is 100 (82 + 18 = 100) and the digit at the hundred's place in both multiplicand and multiplier are the same.

$$7 \; \fbox{82}$$
$$\text{x } 7 \; \fbox{18}$$

LHS = 7 x 8 = 56
RHS = 82 x 18 = 1476
Hence, 782 x 718 = 561476

(NB:– Multiplication of right hand digits 82 x 18 can easily be done by the Vedic sutra, **Urdhva Tiryagbhyam**.

**Example 36:** Multiply 978 by 922

**Solution:** Here the sum of the last two digits of the multiplicand and multiplier is 100 (78 + 22 = 100) and the digit at the hundred's place in both multiplicand and multiplier are same.

$$9 \; \fbox{78}$$
$$\text{x } 9 \; \fbox{22}$$

LHS = 9 x 10 = 90
RHS = 78 x 22 = 1716
Hence, 978 x 922 = 901716

(NB:– Multiplication of right hand digits 78 x 22 can easily be done by the Vedic sutra, **Urdhva Tiryagbhyam**.

**Example 37:** Multiply 811 by 889

**Solution:** Here the sum of the last two digits of the multiplicand and multiplier is 100 (11 + 89 = 100) and the digits at the hundred's place in both multiplicand and multiplier are the same.

$$8 \; \fbox{11}$$
$$\text{x } 8 \; \fbox{89}$$

LHS = 8 x 9 = 72
RHS = 11 x 89 = 979

The RHS should consist of 4 digits, so put one zero before 979, making it 0979

Hence, 811 x 889 = 720979

(NB:– Multiplication of the right hand digits, 11 x 89, can easily be done. Please learn the multiplication by 11 given in the chapter, *Multiplication through observation*).

**Example 38:** Multiply 891 by 809

**Solution:** Here the sum of the last two digit of the multiplicand and multiplier is 100 (91 + 09 = 100) and the digit at the hundred's place in both multiplicand and multiplier are the same.

```
  8 (91)
x 8 (09)
```

LHS = 8 x 9 = 72

RHS = 91 x 9 = 819

The RHS should consist of 4 digits, so put one zero before 819, making it 0819

Hence, 891 x 809 = 720819

## वामनल्यायोह दशकः॰उपि (Vamanlyayoh Dasake Api)

This sutra is not taken from the Vedic Mathematics book written by Jagad guru Bharti Krishna Tirtha ji Maharaj and the original source of this sutra is not known to me. While writing this book, I had gone through more than 50 books and hundreds of websites and I noticed that this sutra serves the purpose of faster calculation and I could not stop myself from introducing this sutra for my avid readers.

This sutra is applicable when the sum of digits placed at the tens' place in the multiplicand and multiplier is 10 and the unit digit of both multiplicand and multiplier is the same.

**Rule:**

- The answer of such questions consists of two parts.
- LHS = Product of two left digits whose sum is 10 + Unit digit
- RHS = Square of Unit digit

Let us take a few examples to understand the modus operandi more clearly.

**Example 39:** Multiply 68 by 48

**Solution:** Here the sum of the tens' digit of multiplicand and multiplier is 10 (6 + 4 = 10).

$$\begin{array}{r} 6\ \ 8 \\ \times\ \ 4\ \ 8 \\ \hline \end{array}$$

LHS = Product of digits at ten's place + unit digit
    = 6 x 4 + 8 = 32
RHS = Square of unit digit
    = 8 x 8 = 64
Hence, 68 x 48 = 3264

**Example 40:** Multiply 72 by 32

**Solution:** Here the sum of tens' digit of multiplicand and multiplier is 10 (7 + 3 = 10).

$$\begin{array}{r} 7\ \ 2 \\ \times\ \ 3\ \ 2 \\ \hline \end{array}$$

LHS = Product of digits at ten's place + unit digit
    = 7 x 3 + 2 = 23
RHS = Square of unit digit
    = 2 x 2 = 04
RHS should contain 2 digits so a zero is added before 4.
Hence, 72 x 32 = 2304

**Example 41**: Multiply 86 by 26?

**Solution**: Here the sum of tens' digit of multiplicand and multiplier is 10 (8 + 2 = 10).

$$\begin{array}{r} \textcircled{8}\ 6 \\ \times\ \textcircled{2}\ 6 \\ \hline \end{array}$$

LHS = Product of digits at ten's place + unit digit
= 8 x 2 + 6 = 22
RHS = Square of unit digit
= 6 x 6 = 36
Hence, 86 x 26 = 2236

**Example 42**: Multiply 97 by 17?

**Solution**: Here the sum of tens' digit of multiplicand and multiplier is 10 (9 + 1 = 10).

$$\begin{array}{r} \textcircled{9}\ 7 \\ \times\ \textcircled{1}\ 7 \\ \hline \end{array}$$

LHS = Product of digits at ten's place + unit digit
= 9 x 1 + 7 = 16
RHS = Square of unit digit
= 7 x 7 = 49
Hence, 97 x 17 = 1649

# वामनल्यायोह दशकः गुणिजःऽपि (Vamanlyayoh Dasake Gunijah Api)

Like the previous sutra, this sutra is not taken from the original Vedic Mathematics book available in the market. The origin of this sutra is not known to me, but again it is one of the interesting sutras that will make our calculations easier and that is my purpose of writing this book.

This sutra is applicable **when the sum of the left two digits, other than the unit, is a multiple of ten and the unit digits are the same.**

**Rule:**

- The result consists of two parts.
- LHS = Product of two left digits whose sum is a multiple of 10 + sub base x unit digit
- RHS = square of unit digit

Let us see its working with the help of these examples.

**Example 43:** Multiply 136 by 76

**Solution:** Here the sum of the left digits in multiplicand and multiplier, other than the unit digit, is 13 + 7 = 20. This sum is absolutely the multiple of 10.

$$20 = \underset{\text{Sub-base}}{2} \quad x \quad \underset{\text{Base}}{10}$$

$$\begin{array}{cc} 13 & 6 \\ x \quad 7 & 6 \end{array}$$

LHS = Product of two left digits whose sum is a multiple of 10 + sub base x unit digit
= 13 x 7 + 2 x 6 = 103
RHS = Square of Unit digit
= 6 x 6 = 36
Hence, 136 x 76 = 10336

**Example 44:** Multiply 278 by 238

**Solution:** Here the sum of the left digits in multiplicand and multiplier, other than the unit digit, is 27 + 23 = 50. This sum is absolutely the multiple of 10.

$$50 = \underset{\text{Sub-base}}{5} \quad x \quad \underset{\text{Base}}{10}$$

$$\begin{array}{r} \fbox{27} \quad 8 \\ \times \quad \fbox{23} \quad 8 \\ \hline \end{array}$$

LHS = Product of two left digits whose sum is a multiple of 10 + sub base x unit digit

= 27 x 23 + 5 x 8 = 661

(The multiplication of 27 x 23 can be done easily by *Antyayoh Dasake api sutra*)

RHS = Square of Unit digit

= 8 x 8 = 64

Hence, 278 x 238 = 66164

**Example 45:** Multiply 459 by 159

**Solution:** Here the sum of the left digits in multiplicand and multiplier, other than the unit digit, is 45 + 15 = 60. This sum is absolutely the multiple of 10.

60 = 6  x   10

Sub-base      Base

$$\begin{array}{r} \fbox{45} \quad 9 \\ \times \quad \fbox{15} \quad 9 \\ \hline \end{array}$$

LHS = Product of two left digits whose sum is a multiple of 10 + sub base x unit digit

= 45 x 15 + 6 x 9 = 729

RHS = Square of Unit digit

= 9 x 9 = 81

Hence, 459 x 159 = 72981

## Urdhva Tiryagbhyam (उर्ध्वतिर्यग्भ्याम्)

So far we have discussed seven Vedic sutras, but all of them have limited applicability. We shall now proceed to deal with a general formula of multiplication which is applicable in all the cases. This sutra is widely known as **Urdhva Tiryag** sutra, which means **"Vertically and Cross-wise"**. Once you get mastery over this method you can multiply a 5 digit multiplication of any number in 15 seconds. Initially this method will seem tough to work out, but believe me; I have seen the change in calculating power of students after learning this sutra. The best feature of this method is:

a) With practice, you can multiply any digits of number and obtain the result in one line.

b) You are free to multiply from both ends due to the flexibility of Vedic Sutra..

A further simplification can be understood by the **Dot and Stick Method..**

## Dot and Stick Method

*Multiplication of 2 digits number*

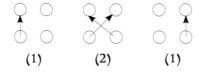

*Multiplication of 3 digits number*

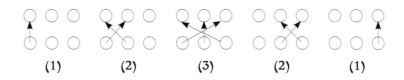

*Multiplication of 4 digits number*

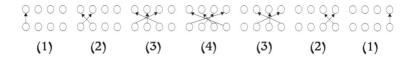

## Multiplication of 5 digits number

(1)    (2)    (3)    (4)    (5)    (4)    (3)    (2)    (1)

The above pictorial representation may be extended upon need. The dot and stick method may bring some amount of discomfort initially, but the more you practice, the more comfortable you will feel with it. At the very beginning you might get puzzled over the multiplication by dot and cross as shown above pictorially, but here is an interesting technique to learn the dot and cross method. See this unique pattern of multiplication:

$$11 \times 11 = 121$$
$$111 \times 111 = 12321$$
$$1111 \times 1111 = 1234321$$
$$11111 \times 11111 = 123454321$$

------------------------------------

------------------------------------

Looking upon the multiplication of 11 x 11, 111 x 111 ...and the dot multiplication shown pictorially, you may find a very special clue. Did you notice the clue?

If you are multiplying a two-digit number, see the product of 11 X 11 i.e. 121 and now notice the number written below the dots of 2 digit multiplication.

Did you find any similarity?

Yes, the same 121 is written there. This process can be summed up in three points.

- First multiply vertically the number placed at unit digits.
- Find the sum of the cross product of the two figures.
- Finally, multiply vertically the remaining digits.

Similarly, if you are multiplying a three-digit number, look at the product of 111 X111 = 12321, this is the same number written at the bottom of the dot multiplication.

Let us take a few examples to clarify the dot and cross product technique.

## A. Multiplication of two-digit numbers

**Example 46:** Multiply 76 by 42

**Solution:**

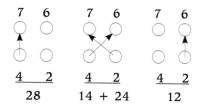

Arranging the number and adding them from right to left, taking only one digit at a time, we get the final result.

= 28 | 38 | 12

= 3192

**Example 47:** 92 X 18 = ?

**Solution:** Arranging the number on the dots.

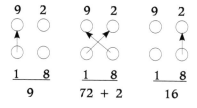

Arranging the number and adding them from right to left, taking only one digit at a time, we get the final result.

= 9 | 74 | 16

= 1656

**Example 48:** Multiply 56 by 34

**Solution:** Arranging the number on the dots.

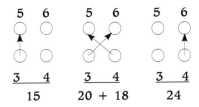

Arranging the number and adding them from right to left, taking only one digit at a time, we get the final result.

$$= 15 \ | \ 38 \ | \ 24$$

$$= 1904$$

Once the concept is clear, the whole process can be done mentally in one line.

**Example 49:** Multiply 77 by 39.

**Solution:** I do hope the concept of dot and cross technique is clearly understood by you. Here is the one liner method. The sum of the cross multiplication of dots in the second stage has to be done mentally.

```
    7        7
  x 3        9
  = 21 |  84  | 63
```

$$= 3003$$

In the 1$^{st}$ vertical separator 7 x 9 = 63 is written, in the second vertical separator the sum of the cross product of 7 x 9 and 3 x 7 is written directly i.e. 7 x 9 + 3 x 7 = 6 3 + 2 1 = 8 4. In the third vertical separator, 7 X 3 = 21 is placed. As told earlier, take only one digit in each separator and add the remaining digit to the next digit separator as shown, in the  direction of the arrow.

## B. Multiplication of three- digit numbers

**Example 50**: Multiply 566 by 281

**Solution**:

Arrange the number on the dots as shown below.

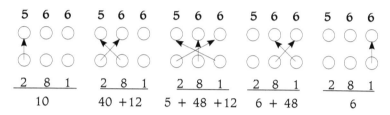

Arrange each product with vertical separator as shown below.

10 | 5 2 | 6 5 | 5 4 | 6

= 159046

**Example 51**: Multiply 659 by 898

**Solution**: Arranging the numbers on the dots.

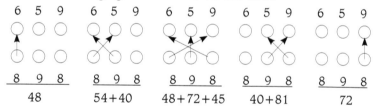

Arranging the product with vertical separator as shown below:

= 4 8 | 9 4 | 1 6 5 | 1 2 1 | 7 2

= 48 | 94 | 165 | 121 + 7 | 2

= 48 | 94 | 165 | 12 8 | 2

$$= 48 \mid 94 \mid 165 + 12 \mid 8 \mid 2$$
$$= 48 \mid 94 \mid \underline{17} \; 7 \mid 8 \mid 2$$

$$= 48 \mid 94 + 17 \mid 7 \mid 8 \mid 2$$
$$= 4\,8 \mid \underline{11}\,1 \mid 7 \mid 8 \mid 2$$

$$= 4\,8 + 1\,1 \mid 1 \mid 7 \mid 8 \mid 2$$
$$= 5\,9 \mid 1 \mid 7 \mid 8 \mid 2$$
$$= 5\,9\,1\,7\,8\,2$$

Hence  6 5 9 x 8 9 8 = 5 9 1 7 8 2

The above steps are written for the sake of readers to understand the concept more vividly, though it is unnecessary to write all these steps. Keep only one thing in your mind that after one stage of operation is over, keep a single digit in each block and move the remaining to the next. Readers are expected to do these operations involving the addition of two or three number mentally.

**Example 52:** Multiply 247 by 989.

**Solution:** The whole operation of dot and cross method is done here in one line.

$$
\begin{array}{cccccc}
 & 2 & 4 & 7 & & \\
 & \text{X } 9 & 8 & 9 & & \\
\hline
{}_1 8 \mid & {}_5 2 \mid & {}_{11} 3 \mid & {}_9 2 \mid & {}_6 3 & \\
 & 8 & 2 & 3 & 2 & 3 \\
+ & 1 & 5 & 11 & 9 & 6 \\
\hline
 & 2 & 4 & 4 & 2 & 8 & 3 \\
\end{array}
$$

In the very beginning, I had mentioned the fact that  once the multiplication of every digit is completed, you have to add the digits from right to left, taking only one digit in each separator. Here the extra digits, leaving the unit digit in each separator, have been written in sub-script so that further addition becomes easy.

**Example 53**: Multiply 467 by 598.

**Solution**:

```
        4   6   7
    X   5   9   8
  ₂0 | ₆6 | ₁₂1 | ₁ ₁1 | ₅6
        0   6   1   1   6
    +   2   6   12  11  5
        2   7   9   2   6   6
```

**Example 54**: Multiply 526 by 43

**Solution**: This is a 3 x 2 digit multiplication so put a zero in front of 43, making it 043 and now apply the above 3 x 3 operation technique. Arrange the numbers on the dots as shown below.

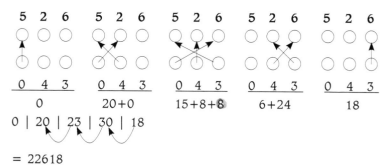

```
5  2  6      5  2  6      5  2  6      5  2  6      5  2  6

0  4  3      0  4  3      0  4  3      0  4  3      0  4  3
   0          20+0        15+8+8        6+24          18

0 | 20 | 23 | 30 | 18
```

= 22618

## C. Multiplication of 4 digits number

The first example here is done in detail with each operation illustrated clearly, but the next example onwards is done in a single line with the hope that readers are well acquainted with the process of multiplication.

**Example 55**: Multiply 6234 by 1235

**Solution**: Arranging the numbers on the dots.

```
6 2 3 4   6 2 3 4   6 2 3 4   6 2 3 4   6 2 3 4   6 2 3 4   6 2 3 4
○ ○ ○ ○   ○ ○ ○ ○   ○ ○ ○ ○   ○ ○ ○ ○   ○ ○ ○ ○   ○ ○ ○ ○   ○ ○ ○ ○
○ ○ ○ ○   ○ ○ ○ ○   ○ ○ ○ ○   ○ ○ ○ ○   ○ ○ ○ ○   ○ ○ ○ ○   ○ ○ ○ ○
1 2 3 5   1 2 3 5   1 2 3 5   1 2 3 5   1 2 3 5   1 2 3 5   1 2 3 5
   6         14        25        46        27        27        20
```

Arrange the numbers in vertical separators.

$$6 \mid 1\,4 \mid 2\,5 \mid 4\,6 \mid 2\,7 \mid 2\,7 \mid 2\,0$$

$$= 7698990$$

**Example 56:** Multiply 8989 with 8892.

**Solution:**

```
    8   9   8   9
  x 8   8   9   2
64 | 1 3 6 | 2 0 8 | 2 3 3 | 1 6 2 | 9 7 | 1 8
```

```
  =   6    4    6    8    3    2   7   8
  +       13   20   23   16    9   1
      7    9    9    3    0    1   8   8
```

**Example 57:** Multiply 2134 by 3261

**Solution:**

```
    2 1 3 4
  x 3 2 6 1
6 | 7 | ₂3 | ₂6 | ₂7 | 4
```
$$= 6958974$$

**Example 58:** Multiply 4382 by 235

**Solution:** This is a 4 x 3 digit multiplication. The 4 x 4 method as explained above can be used here, provided we put a zero in front of 235, making it 0235. Arrangement of dots in this case is shown here.

4 3 8 2   4 3 8 2   4 3 8 2   4 3 8 2   4 3 8 2   4 3 8 2   4 3 8 2

0 2 3 5   0 2 3 5   0 2 3 5   0 2 3 5   0 2 3 5   0 2 3 5   0 2 3 5
   6         8         18        45        43        46        10

= 0 | 8 | 18 | 45 | 43 | 46 | 10

= 1029770

**Example 59**: Multiply 4382 by 35

**Solution**: This is a 4 x 2 digit multiplication, so two zeros have been placed behind 35 to make the calculation by 4 x 4 techniques as described above. Arrangement of dots is shown here for the convenience of readers.

4 3 8 2   4 3 8 2   4 3 8 2   4 3 8 2   4 3 8 2   4 3 8 2   4 3 8 2

0 0 3 5   0 0 3 5   0 0 3 5   0 0 3 5   0 0 3 5   0 0 3 5   0 0 3 5
   0         0         12        29        39        46        10

= 0 | 0 | 12 | 29 | 39 | 46 | 10

= 153370

*Multiplication of 5 digits*

**Example 60**: Multiply 34567 by 12345

**Solution**:

3 4 5 6 7   3 4 5 6 7   3 4 5 6 7   3 4 5 6 7   3 4 5 6 7   3 4 5 6 7   3 4 5 6 7   3 4 5 6 7   3 4 5 6 7

1 2 3 4 5   1 2 3 4 5   1 2 3 4 5   1 2 3 4 5   1 2 3 4 5   1 2 3 4 5   1 2 3 4 5   1 2 3 4 5   1 2 3 4 5

= 3 | 10 | 22 | 40 | 65 | 72 | 70 | 58 | 35

= 426729615

Readers can extend the multiplication of 6 or 7 digits by their own by dot and cross method.

## Multiplication of 3 numbers by Vedic method

The traditional multiplication can't do the multiplication of three numbers in one go, but Vedic Mathematics can, in no time, do the multiplication of three-digit numbers closer to the base and sub-base with ease. For the convenient of readers, the modus operandi of the whole process is illustrated with the help of some examples.

**Case 1: When the numbers to be multiplied are nearer to the power of 10.**

**Example 61:** Multiply 1 2 x 1 3 x 1 5

**Solution:**

Step 1: Write the deviation of each number from its base. Place it against each number. For the above multiplication, the working base = 10, and their deviations from the base are 2, 3 and 5 respectively.

| Number | Deviation from the Base |
|--------|-------------------------|
| 12     | + 2                     |
| 13     | + 3                     |
| 15     | + 5                     |

Step 2: Add all the deviations to the working base.

$10 + 2 + 3 + 5 = 20$

Step 3: Make all possible permutations of the deviation, taking two at a time and add them. For example, the possible permutation of **abc** is **ab, bc and ac.**

$2 \times 3 + 2 \times 5 + 3 \times 5 = 31$

Step 4: Multiply the deviations

$2 \times 3 \times 5 = 30$

Step 5: Arrange the result obtained in above steps as shown here.

20 | 31 | 30

Adding the result from right to left in the direction of the arrows we can find the result.

$$20 \mid 31 \mid 30$$

$$= 2340$$

**Example 62:** Multiply 105 x 104 x 109

**Solution:** The working base = 100

Step 1: Find the deviation of each number from its base. Write it against each number.

| Number | Deviation |
|--------|-----------|
| 105 | + 5 |
| 104 | + 4 |
| 109 | + 9 |

Step 2: Find the sum of base and deviations.

Base + deviations = 100 + 5 + 4 + 9 = 118

Step 3: Multiply the deviations in pairs of two and sum up the results so obtained

5 x 4 + 4 x 9 + 5 x 9 = 101

Step 4: Multiply the deviations.

5 x 4 x 9 = 180

Step 5: Arrange the result of all the above steps in a vertical separator and add them up, from right to left as done previously. But never forget to keep two digits in each separator as constant because the base taken here is 100.

$$118 \mid 101 \mid 180$$

$$= 1190280$$

**Example 63:** 989 x 995 x 1012 = ?

**Solution:** The working base for the above number = 1000

Step 1: Find the deviation of each number from its base. Write it against each number.

| Number | Deviation |
|--------|-----------|
| 989    | $-11$     |
| 995    | $-5$      |
| 1012   | $+12$     |

Step 2: Find the sum of base and deviations.

Base+ deviations $= 1000 - 11 - 5 + 12 = 996$

Step 3: Multiply the deviations in pair of two and sum up the result so obtained.

$(-11) \times (-5) + 5 \times (-12) + (-11) \times (+12)$

$= 55 - 60 - 132 = -137$

Step 4: Multiply the deviations

$-11 \times -5 \times 12 = 660$

Step 5: Arrange the result of all the above steps in a vertical separator.

$= 996 \mid (-137) \mid 660$

$= 995 \mid 1000 - 137 \mid 660$

$= 995 \mid 863 \mid 660$

$= 995863660$

*Case 2: When the working base is 50, 500, 5000...*

The working rule of case 2 is same as that of case 1, except for the following:

- Write the last result in column.
- Keep the right hand figure of column 3 intact and divide the figure in the $2^{nd}$ column by $\frac{1}{2}$ and the figure in the $1^{st}$ column by $\frac{1}{4}$.
- Add the digits of each column as shown in the example.

**Example 64:** Multiply 54 x 56 x 51

**Solution:**

Step 1: Write the deviation of each number from its base. Place it against each number.

| Number | Deviation |
|--------|-----------|
| 54 | + 4 |
| 56 | + 6 |
| 51 | + 1 |

Step 2: Find the sum of base and deviations.

$$50 + 4 + 6 + 1 = 61$$

Step 3: Multiply the deviations in pairs of two and sum up the results so obtained

$$4 \times 6 + 4 \times 1 + 6 \times 1 = 34$$

Step 4: Multiply the deviations

$$4 \times 6 \times 1 = 24$$

Arrange the result obtained above in a vertical separator.

$$61 \mid 34 \mid 24$$

Make three columns and write the result accordingly, as shown here.

| Column 1 | Column 2 | Column 3 |
|----------|----------|----------|
| 61 | 34 | 24 |
| = 61 x ¼ | 34 x ½ | 24 |
| = 15 + ¼ | 17 | 24 |
| = 15 | 25 + 17 | 24 |
| = 15 42 24 | | |

Here the theoretical base = 100 and working base = 50 = 100|2, so ½ = 50 and ¼ = 25 is taken.

**Example 65**: Multiply 54 x 48 x 61

**Solution**: The working base = 50 = 100 | 2 and Theoretical base = 100

Step 1: Find the deviation of each number from its base. Write it against each number.

| Number | Deviation |
|--------|-----------|
| 54 | +4 |
| 48 | –2 |
| 61 | +11 |

Step 2: Find the sum of base and deviations.

Base + deviations = 50+ 4 − 2 + 11 = 63

Step 3: Multiply the deviations in pairs of two and sum up the results so obtained

4 x (− 2) + 4 x 11 + 11 x (− 2) = 14

Step 4: Multiply the deviations.

4 x (− 2) x 11 = − 88

The final result is 63 | 14 | − 88

Write the result in columns.

| Column 1 | Column 2 | Column 3 |
|---|---|---|
| 63 x ¼ | 14 x ½ | −88 |
| = 15 + ¾ | 7 | −88 |
| = 15 | 75 + 7 | −88 |
| = 15 | 82 | −88 |
| = 15 | 81 | 100 −88 |
| = 15 | 81 | 12 |

Hence, 54 x 56 x 61 = 158112

## Multiplication of four numbers by Vedic Method

If you are asked to multiply four numbers the first thing that will strike you is to multiply the first two numbers and then multiply the product with the third number and finally the product with the fourth number. Don't you think that even multiplication of 4 two-digit numbers will take 5 minutes of your time? On the contrary, multiplication of four numbers near the same base or sub-base can be done through the Vedic method quite easily and that too in a few seconds. Are you ready to ride on the chariot of Vedic Sutra?

The answer of such multiplication will consist of 4 parts.

**Rule:**

• First, write the deviation of numbers from its base against the number.

Suppose you have to multiply a x b x c x d if the deviation of these numbers from its base is $d_1$ $d_2$ $d_3$ and $d_4$, then we will write–

| a | $d_1$ |
|---|---|
| b | $d_2$ |
| c | $d_3$ |
| d | $d_4$ |

- $1^{st}$ part = Sum of any number and other three deviations.
  = a + ( $d_2$ + $d_3$ + $d_4$ ) or b + ($d_1$ + $d_3$ + $d_4$) ,or c + ( $d_1$ + $d_2$ + $d_4$) or d + ($d_2$ + $d_3$ + $d_4$)

- $2^{nd}$ part = Sum of the product of two deviations

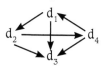

  = $d_1d_2$ + $d_1d_3$ + $d_1d_4$ + $d_2d_3$ + $d_2d_4$ + $d_3d_4$

- $3^{rd}$ part = Sum of the product of three numbers at a time
  = $d_1d_2d_3$ + $d_1d_2d_4$ + $d_1d_3d_4$ + $d_2d_3d_4$

- $4^{th}$ part = Product of all deviations
  = $d_1d_2d_3d_4$

**Example 66:** Multiply 102 x 103 x 104 x 105

**Solution:** Here all the numbers are near the base 100 and their deviations are respectively 2, 3, 4 and 5.

| 102 | + 2 |
|---|---|
| 103 | + 3 |
| 104 | + 4 |
| 105 | + 5 |

- $1^{st}$ part = Sum of any number and other three deviations.
  = 102 + 3 + 4 + 5= 114

- $2^{nd}$ part = Sum of the product of two deviations
  = 2 x 3 + 2 x 4 + 2 x 5 + 3 x 4 + 3 x 5 + 4 x 5 =71

- $3^{rd}$ part = Sum of the product of three numbers at a time
  = 2 x 3 x 4 + 2 x 3 x 5 + 2 x 4 x 5 + 3 x 4 x 5
  = 154

- $4^{th}$ part = Product of all deviations
  = 2 x 3 x 4 x 5= 120
  Hence, 102 x 103 x 104 x 105 = 114 | 71 | 154 | 120

  = 114 | 72 | 55 | 20 = 114725520

Since the base is 100, each part will contain a maximum of two digits and the excess digits will be transferred to the next part.

**Example 67**: Multiply 995 x 996 x 997 x 998

**Solution**: Here all numbers are near the base 100 and their deviations are respectively 2, 3, 4 and 5.

```
995    −5
996    −4
997    −3
998    −2
```

- $1^{st}$ part = Sum of any number and other three deviations.
  = 995 − 4 − 3 − 2 = 986

- $2^{nd}$ part = Sum of the product of two deviations
  = (−5) x (−4) + (−5) x (−3) + (−5 ) x (−2) + (−4) x (−3) + (−4) x (−2) + (−3) x(−2)
  = 20 + 15 + 10 + 12 + 8 + 6 = 71

- $3^{rd}$ part = Sum of the product of three numbers at a time
  = (−5) x (−4) x (−3) + (−5) x (−4) x (−2) + (−4) x (−3) x (−2) + (−5) x (−3) x (−2)
  = − 60 − 40 − 24 − 30
  = −154

- $4^{th}$ part = Product of all deviations
  = − 2 x − 3 x− 4 x− 5= 120

Hence, 995 x 996 x 997 x 998 = 986 | 71 | −154 | 120

Here, the base is taken as 1000, so each part will consist of three digits. In the $2^{nd}$ part, we have a two-digit number, so put a zero in front of it, making it 071. In the $3^{rd}$ part, we have a negative number, so subtract 1 from the $2^{nd}$ part and the negative number of part 3 from the base 1000. Our answer will now look like–

$$
\begin{aligned}
995 \times 996 \times 997 \times 998 &= 986 \mid 071 \mid -154 \mid 120 \\
&= 986 \mid 070 \mid 1000 - 154 \mid 120 \\
&= 986 \mid 070 \mid 846 \mid 120 \\
&= 986070846120
\end{aligned}
$$

*Case 2: When the base is the multiple of 100, 1000... etc.*

Suppose the four numbers to be multiplied are closer to 200, then we shall first write 200 = 2 x 100. Here Base = 100 and sub- base = 2. In such a case, the above process will be applied with a slight change. In the $1^{st}$ part, multiply the result with (sub-base)$^3$, $2^{nd}$ part with (sub-base)$^2$ and $3^{rd}$ part will be multiplied by the sub-base. The number of zeros in the base will decide the number of digits to be kept in each part while writing the final answer.

**Example 68:** Multiply 506 x 507 x 508 x 509

**Solution:**

Here all the numbers to be multiplied are near to 500 = 5 x 100, hence Base = 100 and sub-base = 5. Deviations of the number from 500 are 6, 7, 8 and 9 respectively

$$
\begin{array}{ll}
506 & + 6 \\
507 & + 7 \\
508 & + 8 \\
509 & + 9
\end{array}
$$

- $1^{st}$ part = (Sum of any number and other three deviations) x (Sub-base)$^3$

  = (506+7+8+9) x $5^3$= 530 x 125 = 66250

  (Refer **Multiplication through observation** for multiplying a number by 125)

- $2^{nd}$ part = (Sum of product of two deviations) x (Sub-base)$^2$

  = (6 x 7 + 6 x 8 + 6 x 9 + 7 x 8 + 7 x 9 + 8 x 9) x $5^2$ = 335 x 25 = 8375

  (Refer **Multiplication through observation** for multiplying a number by 25)

- $3^{rd}$ part = (Sum of product of three numbers at a time) x (Sub-base)

  = (6 x 7 x 8 + 6 x 7 x 9 + 6 x 8 x 9 + 7 x 8 x 9 ) x 5 = 1650 x 5 = 8250

- $4^{th}$ part = Product of all deviations

  = 6 x 7 x 8 x 9 = 3024

Hence, 506 x 507 x 508 x 509 = 66250 | $_{83}$75 | $_{82}$50 | $_{30}$24

Since the base = 100, each part will contain maximum two digits and the rest will be transferred to the next part. The excess digit is written in subscript.

  506 x 507 x 508 x 509 = 66250 | $_{83}$75 | $_{82}$50 | $_{30}$24

  = 66334578024

# 4

# Multiplication through Observation

## Introduction

The best part of Vedic maths is that you can do calculations in a few seconds and sometimes, orally. In this chapter we shall learn how to do quick and flawless calculations and that too by mere observation. Once the techniques for such special cases are mastered, you will feel enthusiastic enough to do some more tricks of your own.

## Mental Multiplication

### A: Multiplication by 11

Multiplication of any number with 11 can be done orally in a single line. Once the technique for multiplication of a number with 11 is mastered, it can be further extended for a number such as 22, 33, 44 etc by simply splitting the multiplicand as 11 x 2, 11 x 3 or 11 x 4. In mensuration, you need to calculate the volume and surface area of three-dimensional objects such as cylinder, sphere, cone, pyramid, frustum etc and there you need to multiply the number by 22 ($\pi$ =22/7). This method will help you immensely here.

### Rule:

- Place the number to be multiplied by 11 in a bracket and put zeros on either side.

- Start adding the two numbers at a time from right to left. If the sum of two numbers in any case exceeds 10, the digit at the tenth place shall be carried over to the next sum, as is usually done in simple addition.

**Example 1:** Multiply 3251 by 11

**Solution:**

Place the number in a bracket and put zeros on either side.

Add the digit from the right to left as shown above.

0+ 3 | 3+2 | 2 + 5 | 5 + 1 | 1 + 0

= 3 5 7 6 1

Hence 3251 x 11 = 35761

**Example 2:** Multiply 4876254 by 11

**Solution:**

Place the number in a bracket and put zeros on either side.

0 ( 4  8  7  6  2  5  4 ) 0

Add the digit from the right to left as shown above.

= 0+4| 4+8 | 8+7 | 7+6 | 6+2 | 2+5 | 5+4 | 4+0

= 4 | 12 | 15 | 13 | 8 | 7 | 9 | 4

= 4 | 12 | 16 | 3 | 8 | 7 | 9 | 4

= 4 | 13 | 6 | 3 | 8 | 7 | 9 | 4

= 5 3 6 3 8 7 9 4

Hence, 4876254 x 11 = 53638794

**Example:** Multiply 384 by 11

**Solution:** Place the number in a bracket and put zeros on either side.

$$0 \left( \; 3 \quad 8 \quad 4 \; \right) 0$$

Add the digit from the right to left as shown above.

$$0+3 \mid 3+8 \mid 8+4 \mid 4+0$$

$$= 3 \mid 11 \mid 12 \mid 4$$

$$= 3 \mid 12 \mid 2 \mid 4$$

$$= 4 \; 2 \; 2 \; 4$$

Hence, 384 x 11 = 4224

**B: Multiplication by 111**

Rule:- Multiplication with 111 involves the same pattern with a slight difference.

• Place the number to be multiplied inside a bracket and two zeros on either sides of the bracket.

• Add from right to left, taking the sum of three digits at a time.

• If the sum of the digits exceeds 10, the digit at the ten's place will be carried over to the next sum.

**Example:** Multiply 34 by 111

**Solution:** Place the number in a bracket and put 2 zeros on either side.

Keep adding three digits from the right at a time as shown above.

$$= 0 + 0 + 3 \mid 0 + 3 + 4 \mid 3 + 4 + 0 \mid 4 + 0 + 0$$

$$= 3774$$

**Example 2:** Multiply 497 by 111

**Solution:** Place the number in a bracket and put zeros on either side.

$$0 \ 0 \left( 4 \quad 9 \quad 7 \right) 0 \ 0$$

Keep adding from right to left, taking the sum of three digits at a time.

$$0 \ 0 \left( 4 \quad 9 \quad 7 \right) 0 \ 0$$

= 0+0+4 | 0+4+9 | 4+9+7 | 9+7+0 | 7+0+0

= 4 | 13 | 20 | 16 | 7

= 4 | 13 | 21 | 6 | 7

= 4 | 15 | 1 | 6 | 7

= 5 5 1 6 7

**C: Multiplication with 1111**

Rule: Multiplication with 1111 involves the same sort of operations as discussed above.

*   Place the number to be multiplied inside a bracket and three zeros on either side of the bracket.
*   Add from right to left, taking the sum of four digits at a time.
*   If the sum of the digits exceeds 10, the digit at the ten's place will be carried over to the next sum.

**Example:** Multiply 2172 by 1111

**Solution:** Place the number in a bracket and put 3 zeros on either side.

$$0 \quad 0 \quad 0 \left( 2 \quad 1 \quad 7 \quad 2 \right) 0 \quad 0 \quad 0$$

Keep adding from right to left, taking 4 digits at a time.

Step 1:   0   0   0   (2   1   7   2)   0   0   0

$$0 + 0 + 0 + 2 = 2$$

Step 2:   0   0   0   (2   1   7   2)   0   0   0

$$0 + 0 + 2 + 7 = 9$$

Step 3:   0   0   0   (2   1   7   2)   0   0   0

$$0 + 2 + 7 + 1 = 10$$

Step 4 :   0   0   0   (2   1   7   2)   0   0   0

$$2 + 7 + 1 + 2 = 12$$

Step 5:   0   0   0   (2   1   7   2)   0   0   0

$$0 + 2 + 1 + 7 = 10$$

Step 6:   0   0   0   (2   1   7   2)   0   0   0

$$0 + 0 + 2 + 1 = 3$$

Step 7:   0   0   0   (2   1   7   2)   0   0   0

$$0 + 0 + 0 + 2 = 2$$

Arranging all the steps in one line, we get—

$$= 2 \mid 3 \mid 10 \mid 12 \mid 10 \mid 9 \mid 2$$

$$= 2 \mid 3 \mid 10 \mid 13 \mid 0 \mid 9 \mid 2$$

$$= 2 \mid 3 \mid 11 \mid 13 \mid 0 \mid 9 \mid 2$$

$$= 2 \mid 4 \mid 1 \mid 3 \mid 0 \mid 9 \mid 2$$

Hence, 2172 x 1111 = 2413092

## D: Multiplication by 25

Rule: This is a very simple technique. When you are multiplying a number by 25, put two zeros (00) to the right of the multiplicand and divide it by 4.

**Example 1:** Multiply 16 by 25

**Solution:** Put two zeros to the right of 16, i.e. 1600
Divide it by 4 = 1600 / 4 = 400
25 x 16 = 400

**Example 2:** Multiply 98 by 25

**Solution:** Put two zeros to the right making it 9800
Divide it by 4, 9800 / 4 = 2450
98 x 25 = 2450

**Example 3:** Multiply 428764 by 25

**Solution:** Put two zeros to the right of 428764, i.e. 42876400
Divide it by 4 = 42876400 / 4 = 10719100
428764 x 25 = 10719100

**Example 4:** Multiply 82456 by 25

**Solution:** Put two zeros to the right of 82456, i.e. 8245600
Divide it by 4 = 8245600 / 4 = 2061400
82456 x 25 = 2061400

## E: Multiplication by 125

Rule: Put three zeros to the right of the multiplicand and divide it by 8.

**Example 1:** Multiply 624 by 125

**Solution:** Place 3 zeros after 624, making it 624000.
Divide it by 8= 624000/8 = 78000
Hence, 624 x 125 = 78000

**Example 2:** Multiply 48 by 125

**Solution:**    Place 3 zeros after 48, making it 48000
                 Divide it by 8= 48000/8 = 6000
                 Hence, 48 x 125 = 6000

**Example 3:** Multiply 24376 by 125

**Solution:**    Place 3 zeros after 24376, making it 24376000
                 Divide it by 8= 24376000/8 = 3047000
                 Hence, 24376 x 125 = 3047000

**F: Multiplying with 625**

Rule: Put 4 zeros (0000) to the right of the number and divide it by 16.

**Example 1:** Multiply 428 by 625

**Solution:**    Place 4 zeros after 428, making it 4280000
                 Divide it by 16 = 4280000/16 = 267500
                 Hence, 428 x 625 = 267500

**Example 2:** Multiply 246284 by 625

**Solution:**    Place 4 zeros after 246284, making it 2462840000
                 Divide it by 16 = 2462840000/16 = 153927500
                 Hence, 246284 x 625 = 153927500

**Example 3:** Multiply 144 by 625

**Solution:**    Place 4 zeros after 144, making it 1440000
                 Divide it by 16 =1440000/16 = 90000
                 Hence, 144 x 625 = 90000

**G: Multiply by 5**

Rule: Put one zero to the right of the number and divide it by 2.

**Example 1:** Multiply 42 by 5

**Solution:**    Place 1 zero after 42, making it 420

Divide it by 2 = 420/2 = 210

Hence, 42 x 5 = 210

**Example 2:** Multiply 5986 by 5

**Solution:** Place 1 zero after 5986, making it 59860

Divide it by 2 = 59860/2 = 29930

Hence, 5986 x 625 = 29930

**H: Multiply by 50**

Rule: Put two zeros to the right of the number and divide it by 2.

**Example 1:** Multiply 47 by 50

**Solution:** Place 2 zero after 47, making it 4700

Divide it by 2 = 4700/2 = 2350

Hence, 47 x 50 = 2350

**Example 2:** Multiply 62876 by 50

**Solution:** Place 2 zeros after 62876, making it 6287600

Divide it by 2 = 6287600/2 = 3143800

Hence, 62876 x 50 = 3143800

**I: When the sum of the unit's place digit is 10 and the rest of the digits are the same**

**Rule:**

- Multiply the unit digit whose sum is 10 and place it on the right side.
- Increase the multiplicand by 1 and then multiply it with the original multiplier. Put this result to the left.

**Example:** Multiply 46 by 44

**Solution:** Here the sum of the unit digit is 10 and the digit at the ten's place in the multiplier and multiplicand are also same.

```
  4 6
X 4 4
```

Multiply the encircled digit and write it to the right side.

$$4\,\boxed{6}$$
$$\text{x } 4\,\boxed{4}$$
$$\overline{\phantom{xx}/\ 24}$$

Increase the multiplicand by 1 and then multiply it with the original multiplier.

$$(4+1)=\quad \boxed{5}\ 6$$
$$\text{x}\ \boxed{4}\ 4$$
$$\overline{20/\ 24}$$

Hence, 46 x 44 = 2024

**Example:** Multiply 113 by 117

Solution: Here the sum of the unit digit is 10 and the digit at the ten's place in the multiplier and multiplicand are also same.

$$11\ \boxed{3}$$
$$\text{x } 11\ \boxed{7}$$

Multiply the encircled digit and write it to the right side.

$$11\ \boxed{3}$$
$$\text{x } 11\ \boxed{7}$$
$$\overline{\phantom{xx}/\ 21}$$

Increase the multiplicand by 1 and then multiply it with the original multiplier.

$$(11+1)\ =\quad \boxed{12}\ 3$$
$$\text{x } 11\ 7$$
$$\overline{132/\ 21}$$

Hence, 113 x 117 = 13221
(Multiplication of 12 and 11 can be done as discussed in part A in this chapter itself.)

**Example:** Multiply 168 by162

**Solution:** Here the sum of the unit digit is 10 and the digit at the ten's place in the multiplier and multiplicand are also same.

$$16\,8$$
$$\times\ 16\,2$$

Multiply the encircled digit and write it to the right side.

$$16\ 8$$
$$\times\ 16\ 2$$
$$/16$$

Increase the multiplicand by 1 and then multiply it with the original multiplier.

$$(16+1)\ =\quad 17\ 8$$
$$\times\ 16\ 2$$
$$272/\ 16$$

Hence, 168 x 162 = 27216

(The multiplication of 17 x 16 can be done through the Vedic sutra Urdhyagtiryagbhyam)

**J: When the sum of the last two digits is 100 and the rest of the digits at the hundred's places are the same).**

This method is applicable when the sum of the unit's and ten's place digits of the multiplicand and multiplier is 100 and hundred's place digit is the same.

$$H\ T\ O$$
$$1\ 0\ 3$$
$$\times\ 1\ 9\ 7$$

**Example:** Multiply 103 by 197

**Rule:**

• Multiply the number (unit and ten's place digit) of the multiplicand and multiplier whose sum is 100. The product should be of 4 digits. If the product in RHS is less than four digits, place zero at the extreme left.

```
H T O
1/0 3
X 1\9 7/
 /0291
```

(Since 03 x 97 = 297 contains only 3 digits, one zero has been placed at the extreme left, making it 0297. As explained above, the right side of the product should contain 4 digits. )

*   Increase the multiplicand (digit at hundred's place) by 1 and multiply it with the number placed at the hundred's place in the multiplier.

```
                H T O
(1 + 1)= 2/0 3
X           1\9 7/
        2 /0291
```

Hence, 103 x 197 = 20291

**Example:** Multiply 425 by 475

**Solution:**

Multiply the number (unit and ten's place digit) of the multiplicand and the multiplier whose sum is 100.

```
H T O
4/2 5
X 4\7 5/
 /1875
```

(Here the sum of the encircled number is 100 (25 +75 =100). For multiplication of 25 x 75, see Rule D.

*   Increase the multiplicand (digit at hundred's place) by 1 and multiply it with the number placed at the hundred's place in the multiplier.

$$
\begin{array}{r}
\text{H T O} \\
(4 + 1)= 5\,\boxed{2\ 5} \\
\text{X} \qquad 4\ 7\ 5 \\
\hline
2\ 0\ /1875
\end{array}
$$

Hence, 425 x 475 =201875

**Example:** Multiply 211 by 289

**Solution:**

Multiply the number (unit and ten's place digit) of the multiplicand and the multiplier whose sum is 100.

$$
\begin{array}{r}
\text{H T O} \\
2\,\boxed{1\ 1} \\
\text{X } 4\ 8\ 9 \\
\hline
/0979
\end{array}
$$

[Here the sum of the encircled numbers is 100 (25 +75 =100). For multiplication of 25 x 75, see Rule A. ]

- Increase the multiplicand (digit at the hundred's place) by 1 and multiply it with the number placed at the hundred's place in the multiplier.

$$
\begin{array}{r}
\text{H T O} \\
(2 + 1) = 3\,\boxed{1\ 1} \\
\text{X} \qquad 2\ 8\ 9 \\
\hline
6\ /0979
\end{array}
$$

Hence, 2 1 1 x 2 8 9 = 60979

**K: Multiply any number by 51**

**Rule:**

- Place the multiplicand on the right side. The right side should contain only two digits. If the number of digits at the right side is more than 2, the left- most digit will be carried to the next column on the left side.
- On the left, place half of the multiplicand.
- In case the half of the multiplicand is fractional, add 50 to the digit placed on the right side. This case will occur if the **digit of the multiplicand is odd.**

**Example:** Multiply 42 by 51

**Solution:**          42

                     x 51

- Place 42 at the right side

                     42

                     x 51

                     /42

- Place half of the multiplicand to the left, i.e. 42/2 = 21

                     42

                     x 51

                  21 /42

Hence, 42 x 51 = 2142

**Example:** Multiply 124 by 51

**Solution:**          124

                     x 51

- Place 124 at the right side

                     124

                     x 51

                     /124

- Place half of the multiplicand to the left, i.e. 124/2 = 62. The excess digit from the right side will get transferred to the left side.

                     124

                     x 51

                  62/124

       = 63/24

Hence, 124 x 51 = 6324

**Example:** Multiply 41 by 51

**Solution:**          41

                     x 51

- Place 41 at the right side

    41
    x 51
    /41

- Place half of the multiplicand to the left, i.e. 41/2 = 20½.

    41
    x 51
    20 ½ /41

= 20 / 50 + 41 = 91

Since, the left most part is fractional, add 50 to the right side and remove the fractional part from the left.

Hence, 41 x 51 = 2091

**Example:** Multiply 23 by 51

**Solution:**    23
    x 51

- Place 23 at the right side

    23
    x 51
    /23

- Place half of the multiplicand to the left, i.e. 23/2 = 11½.

    23
    x 51
    11½/23

= 11 / 50 + 23 = 73

Since, the left most part is fractional, add 50 to the right side and remove the fractional part from the left.

Hence, 23 x 51 = 1173

# 5

# Multiplication in Algebra

## Introduction

In the Multiplication chapter, we had discussed various Vedic sutras of multiplying two-digit numbers and three digits number under different conditions.

The Urdhva Tiryagbhyam उर्ध्वतिर्यग्भयाम् formula is also applicable to multiplication of algebraic equations. The present chapter will deal with the following types of equations:–

- Multiplication of Binomials
- Multiplication of Polynomials with equal number of terms
- Multiplication of Polynomials with unequal number of terms.

## Vedic Method of Multiplying Polynomials

In the present chapter, I will just focus on Urdhva Tiryagbhyam (उर्ध्वतिर्यग्भयाम्) formula. The meaning and its working has been explicitly discussed in the multiplication chapter. The Urdhva Tiryagbhyam (उर्ध्वतिर्यग्भयाम्) formula literally means, "Vertical and Cross- wise multiplication". Let me remind you of its working through one example.

### Multiplication of Binomial equations

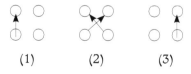

(1)          (2)          (3)

## Multiplication of Trinomials

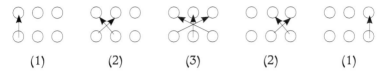

(1)  (2)  (3)  (2)  (1)

The above arrows will help you to recall what you have done previously and ease the understanding process even in Algebraic Multiplication.

*Case 1: Multiplication of Binomial equations.*

**Example:** Multiply  x + 2y and 3x + 4y

Let us first understand it in the traditional way.

$$x + 2y$$
$$\underline{x \quad 3x + 4y}$$
$$= 4y\ (x + 2y) + 3x\ (x + 2y) \qquad \text{(Distributive law)}$$
$$= 4xy + 8\ y^2 + 3\ x^2 + 6\ xy$$
$$= 3\ x^2 + 10\ xy + 8\ y^2$$

## Vedic Method

Before I proceed with the Vedic way, let me explain to you the steps involved in solving.

Step 1: First, write the two variables on the top and put the coefficients from each equation below. Apply the *Urdhya tiryagbhyam* Vedic sutra. **Remember, the carry-over to the preceding column as done in multiplication will not be executed here.**

$$
\begin{array}{cc}
x & y \\
1 & 2 \\
3 & 4 \\
\end{array}
$$

3 |  4+6  | 8

Vertical | sum of cross wise | vertical

1 x 3 | 1 x 4 + 2 x 3 | 2 x 4

3 | 10 | 8

Step 2: Starting from the right, add the variables to these coefficients in the following manner.

(a) The vertical multiplication of y x y is 8. So add $y^2$ to 8 as this is the coefficient of $y^2$.
(b) The crosswise multiplication of x and y , yields result 10, so add xy to 10.
(c) The vertical multiplication of x and x yields 3 as a result so this is the coefficient of $x^2$.
Hence our answer is $3x^2 + 10xy + 8y^2$

**Example:** Multiply  (8x – 3 y)  by (2x + 4y)

**Solution:**

Step 1: Write the two variables on top and their coefficients along their respective signs below them.

Step 2: Do vertical and crosswise multiplication from the right. The multiplication will have the following steps–

a)  4 x – 3 = – 12   (Vertical product)
b)  8 x 4 + 2 x ( – 3 ) = 32 – 6 = 26 (sum of cross wise multiplication)
c)  8 x 2 = 16   (Vertical product of extreme left)

16  |  26  |  –12

Step 3: Place the respective variables to the above coefficient, i.e.

Place $y^2$ to – 12
xy to 26
and $x^2$  to 16.

Hence the result is–

$(8 \, x - 3y) \times (2 \, x + 4 \, y) = 16 \, x^2 + 26 \, xy - 16 \, y^2$

**Example 3:** Multiply $3x - 7 \, y$ by $2 \, x - 5 \, y$

**Solution:** The multiplication of Vedic involves the following steps.

Step 1: Write down the variable on the top and their respective coefficient below, with the proper sign.

Step 2: Do vertical and cross wise multiplication from the right and place the proper variables to the coefficients.

(a) $-7 \, x - 5 = 35$ (Vertical product and coefficient of $y^2$)

(b) $3 \, x - 5 + 2 \, x - 7 = -29$ (Crosswise product and coefficient of xy)

(c) $3 \times 2 = 6$ (Vertical Product of extreme left and coefficient of $x^2$)

Hence the product is–

$$(3x - 7 \, y) \times (2 \, x - 5 \, y) = 6 \, x^2 - 29 \, xy + 35 \, y^2$$

Let me remind you the rule before writing the one line answer to the next few examples that can be done with a little practice.

**Rule:**

- Write down the variable at the top
- Place the coefficients below the variable with the respective sign (+ or −).
- Multiply the number by the vertical and cross-wise method as shown above and also discussed in the Multiplication chapter. Remember one thing – don't take forward any carry-over as done in usual multiplication.
- Place the respective variables to the coefficients to get the one line answer.

Let me take you through a interesting tour of algebraic multiplication on a ride on the very interesting Vedic Chariot called – उर्ध्वतिर्यग्भ्याम् (Urdhva Tiryagbhyam).

**Example:** Multiply ( 2 x + 3 y )by ( 11x + 5 y)

**Solution:**

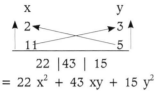

$$22 \mid 43 \mid 15$$
$$= 22 x^2 + 43 xy + 15 y^2$$

**Example:** Multiply 2x – 5 y and 3 x – 7 y

**Solution:**

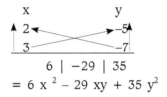

$$6 \mid -29 \mid 35$$
$$= 6 x^2 - 29 xy + 35 y^2$$

**Example:** Multiply $2y^2 + 4z$ by $3y^2 – 7z$

**Solution:**

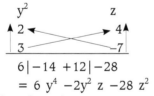

$$6 \mid -14 +12 \mid -28$$
$$= 6 y^4 - 2y^2 z - 28 z^2$$

Since $y^2 \times y^2 = y^4$, 6 is obviously the coefficient of $y^4$. Moreover, the crosswise multiplication gives the coefficient – 2 and it is the coefficient of $y^2z$. The extreme right vertical multiplication – 28 is the coefficient of z x z = $z^2$.

## Multiplication of Algebraic Polynomial

What is a polynomial?

The general equation for a polynomial in the variable $x$ is $a_0 x^n + a_1 x^{n-1} + a_2 x^{n-2} + \ldots$ where $a_0$, $a_1$, etc., are constants and $n$ is the highest power of $x$, called the *degree* of the polynomial.

The multiplication of Algebraic polynomial by the traditional method taught in our classrooms is a little bit confusing and one needs to club variables of the same degree before coming to an answer. Let us take an example and try to solve it by the traditional method.

Multiply $x^2 + 2x + 3$ by $3 x^2 + 2x + 4$

Solution: $(x^2 + 2x + 3) \times (3 x^2 + 2x + 4)$

$= x^2 (3 x^2 + 2x + 4) + 2x (3 x^2 + 2x + 4) + 3 (3 x^2 + 2x + 4)$

[Distributive Property]

$= 3 x^4 + 2 x^3 + 4 x^2 + 6 x^3 + 4 x^2 + 8x + 9 x^2 + 6 x + 12$

$= 3 x^4 + 8 x^3 + 17 x^2 + 14 x + 12$

[on clubbing the power of like terms]

Vedic Method Urdhvatiryagabhyam is equally applicable to the polynomial involving different power of x.

**Rule:**

1. Write the variable above the horizontal line and then list the coefficients of the variable with respective signs below them.

2. Perform the vertical and cross-wise multiplication from the coefficients.

3. Add the variables to the coefficient in respective order.

**Example:** Multiply $(x^2 + 2x + 3)$ by $(3x^2 + 2x + 4)$

**Solution:** This polynomial of $2^{nd}$ degree involves two variables, namely $x^2$, x and one independent term. Place $x^0$ for the independent term. Write the variables at the top and put down the coefficient below the variable shown here.

| $x^2$ | $x$ | $x^0$ |
|---|---|---|
| 1 | 2 | 3 |
| 3 | 2 | 4 |

Let me remind you about the three-digit multiplication with this example to assist you in understanding the concept more comfortably.

**Example:** Multiply 548 by 159.

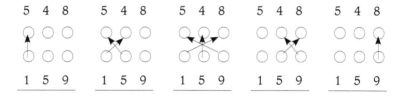

I am hopeful that the above arrow diagram as discussed in the multiplication chapter under Urdhva Tiryagbhyam sutra will help you a lot in understanding the algebraic multiplication effortlessly. I am also listing the step by step procedure for easy understanding.

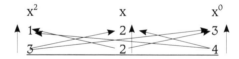

Step 1: Do the vertical and cross-wise multiplication from the right as shown in the above diagram. The method has been explicitly explained below.

    a)  3 x 4 = 12
    b)  2 x 4 + 2 x 3 = 14
    c)  1 x 4 + 3 x 3 + 2 x 2 = 17
    d)  1 x 2 + 3 x 2 = 8
    e)  1 x 3 = 3

Step 2: Place these coefficients below in the answer line column.

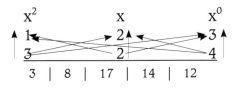

$$3 \mid 8 \mid 17 \mid 14 \mid 12$$

Starting from the extreme right; place the variable to each coefficient in increasing order, i.e. Place $x^0$ to 12, $x^1$ to 14, $x^2$ to 17, $x^3$ to 8 and $x^4$ to 3.

Hence, $(x^2 + 2x + 3) \times (3x^2 + 2x + 4) = 3x^4 + 8x^3 + 17x^2 + 14x + 12$

**Example:** Multiply $2x^2 - 4x + 6$ by $3x^2 - 7x - 2$

**Solution:**

Step 1: Put the variable $x^2$, x and $x^0$ at the top and place the coefficient of the polynomials below it along with the respective sign.

| $x^2$ | x | $x^0$ |
|---|---|---|
| 2 | −4 | 6 |
| 3 | −7 | −2 |

Step 2: Do the vertical and cross-wise multiplication from the right. The steps are shown here for clarity and better understanding.

|  | $x^2$ | x | $x^0$ |
|---|---|---|---|
|  | 2 | −4 | 6 |
|  | 3 | −7 | −2 |

a) $6 \times -2 = -12$

b) $-4 \times -2 + 6 \times -7 = -34$

c) $2 \times -2 + 3 \times 6 + (-4) \times (-7) = 42$

d) $2 \times -7 + 3 \times -4 = -26$

e) $2 \times 3 = 6$

Step 3: Place the coefficient obtained in step 2 below in the answer line. Starting from the extreme right i.e (–12), place the variable to each successive coefficient in increasing order.

The variables placed to the coefficients are shown here.

$x^0$ or no variable to – 12

x to – 34

$x^2$ to 42

$x^3$ to – 26

$x^4$ to 6

Hence ( $2x^2 – 4x + 6$ ) x ( $3 x^2 – 7 x – 2$) = $6 x^4 – 26 x^3 + 42 x^2 – 34 x – 12$

## Case 3: Multiplication of equations having unequal number of terms.

**Rule:** While multiplying two polynomials with unequal terms, we follow the same method discussed so far, with only a simple difference that we place a zero as the coefficients for the variables missing. Let us understand this with an example.

**Example:** Multiply $7x^2 + 6 x + 5$ by x + 9

Since the second equation involves only two terms i.e. x and 9 and the first equation has three terms, namely 7 $x^2$, 6x and 5, this gives us a clue that the $x^2$ term is missing in the second equation. Hence we shall write the second equation as –

x + 9 = 0 $x^2$ + x + 9

Now the multiplication of the above equations will be done like the previous case.

Step 1: Put the variable $x^2$, x and $x^0$ at the top and the coefficient of the variable with respective sign below it.

$$
\begin{array}{ccc}
x^2 & x & x^0 \\
7 & 6 & 5 \\
0 & 1 & 9
\end{array}
$$

Step 2: Do the vertical and cross-wise multiplication.

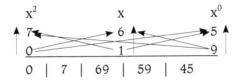

0 | 7 | 69 | 59 | 45

Step 3: Place the respective sign and start placing variables from right to left, increasing the power of the variables in each preceding term,

i.e.  $x^0$ to 45,
    $x^1$ to 59,
    $x^2$ to 69
and  $x^3$ to 7.

Hence, $(7x^2 + 6 x + 5) \times (x + 9) = 7 x^3 + 69 x^2 + 59 x + 45$

**Example:** Multiply $2 x^2 - 7$ by $3 x^2 + 4 x$

Step 1: Place the variables and their corresponding coefficients with respective sign as shown here.

$$
\begin{array}{ccc}
x^2 & x & x^0 \\
2 & 0 & -7 \\
\hline
3 & 4 & 0
\end{array}
$$

Step 2: Do the vertical and crosswise multiplication.

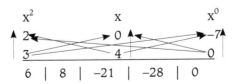

6 | 8 | −21 | −28 | 0

Step 3: Place the variables in ascending order from the right to left.

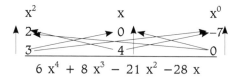

$$6 x^4 + 8 x^3 - 21 x^2 - 28 x$$

Hence, $(2 x^2 - 7) \times (3 x^2 + 4 x) = 6 x^4 + 8 x^3 - 21 x^2 - 28 x$

# 6

# Division

## Introduction

Division is the reverse process of multiplication. If we say $12 \div 3 = 4$, it means $3 \times 4 = 12$. Division basically means repeated subtraction. Let us take an example: $12 \div 3 = ?$

$1^{st}$ stage $= 12 - 3 = 9$
$2^{nd}$ stage $= 9 - 3 = 6$
$3^{rd}$ stage $= 6 - 3 = 3$
$4^{th}$ stage $= 3 - 3 = 0$

Since at the $4^{th}$ stage, we get a remainder of 0, $12 \div 3 = 4$. The conventional method is lengthy and time-consuming as it is based on a hit and trial method. This method is effective as long as the divisor is a single digit, but when the divisor is a bigger number we are at loss because we keep multiplying the dividend by different numbers from 1 to 9. On the other hand, in the Vedic system, we get the quotient and remainder in one line.

## Conventional Method

```
54 ) 4 3 8 5 4 ( 8 1 2        Rough area
     -4 3 2                    54 x 6 = 324
         6 5                   54 x 7 = 378
        -5 4                   54 x 8 = 432
         1 1 4
       - 1 0 8
             6
```

## Vedic one-line Dhwajak method

$$\frac{5^4 \left| \begin{array}{ccccc} 4 & 3 & {}_3 8 & {}_1 5 & {}_1 4 \end{array} \right.}{\left| \begin{array}{ccc} 8 & 1 & 2 \end{array} \right| 6}$$

The conventional method is a fixed stereotypical method where there is no room for experiment, while there are various simple methods of division in Vedic mathematics, where readers have a choice of the best method fitted to that very situation.

In this chapter we shall deal with three Vedic sutras.

In the Nikhilam Method, there is no subtraction to be done at all. The second method, Paravartya Yojyet is based on simple algebraic division and the last method popularly known as Dhwajanka method of Division is based on the long established Vedic process of mathematical calculations. The Dhwajank method is popularly known as Urdhva Tiryagbhyam. Different from the NIkhilam and Paravartya methods, the Dhwajanka method is capable of immediate application to all cases and can be described as the **"crowning gem of all"**for the universality of its application.

## Vedic Method for Division

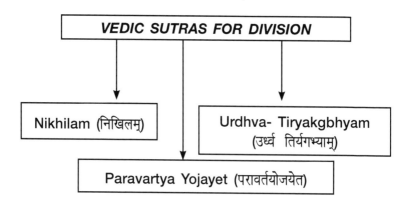

104  *The Essentials of Vedic Mathematics*

## Meaning of Vedic Sutra:

**Nikhiam** (निखिलम्): You must be acquainted with this sutra, Nikhilam Navataha Charmam Dastah. We have used this Vedic sutra in subtraction and multiplication earlier. The meaning of this sutra has been explained in Subtraction – "All from nine and last from 10". This sutra has limited application and is useful when every digit of the divisor is greater than 5. The best part of this sutra is that there is no subtraction process to be carried out at all.

**Paravartya Yojyet** (परावर्तयोजयेत्): This sutra has an upper hand in comparison with the Nikhilam method. The basic meaning of this sutra is "Transpose and Apply." The well-known rule relating to transposition is to invariably change the sign with every change of side. This method like the Nikhilam method has limited application.

**Urdhya-Tiryag bhyam** (उर्ध्वतिर्यगभ्याम्): This Vedic sutra is called the crowning gem of all due to its applicability in all types of division. The meaning is self-explanatory as this method is discussed well in detail in the chapter on Multiplication. This is also called the Dhwajank method of division. The word "dhwaj" means flag and "anka" means digit. Hence, the literal meaning of the word "Dhwajank" is to place a digit or two of the divisor at the top and division is then carried out with the remaining digits of the divisor.

## The Nikhilam method

### Working Rule

1. Take a base (in the power of 10) nearest to the divisor and write its complement below the original divisor, in the divisor column. Complement = Base – Divisor. The complement can be carried out easily by the Nikhilam

Navataha Charmam Dastah method as explained in the Subtraction chapter. Write the complement of the divisor below it.

2. Separate the extreme right digit of the dividend by drawing a slash equal to the number of digits in the divisor. This block is known as the remainder block and the left block is known as the quotient block.

3. The number of digits to be placed in the remainder column should be equal to the number of zeros in the base.

4. Carry down the first digit of divisor in the first nd column. This gives you the first digit of the quotient. Multiply the quotient digit by the complement, and place it in the dividend column; next to the first digit of the dividend.

5. Write mechanically the sum of the digits of the second column to get the second digit of the quotient.

6. Repeat the process until you get a number in the remainder column. If the remainder is greater than the divisor, continue the same process in the remainder block until the digit in the reminder column is less than that of the original divisor.

*Case 1: When the remainder is less than the divisor.*

**Example 1**: Divide 22 by 8

**Solution**:

$$
\begin{array}{r}
\text{Divisor} \quad 8) \; 2 \, / \, 2 \\
\text{Complement} \; 2 \quad \downarrow \underline{\; 4 \;} \quad \text{(Quotient x Complement)} \\
2 \, / \, 6
\end{array}
$$

Here the divisor is nearer to the base 10.

Complement $= 10 - 8 = 2$

- Since the base has one zero, one digit will be separated by a slash for the remainder column. In the above example, the right-most 2 is separated by a slash.

- Carry down the first digit (2) of the dividend. This is the first digit of the quotient.

- Multiply the first digit of the quotient by the complement and place it in the dividend column, next to the first digit of the dividend. Here the first digit of the quotient is 2 and the complement is also 2 hence 2 x 2 = 4 is placed below 2 as shown above.
- Write the sum of the two digits of the remainder column. Since the remainder 6 is less than the divisor,we have completed the division.
- Hence, Quotient = 2 ; Remainder = 6

**Example 2:** Divide 213 by 9

**Solution:**

```
Divisor      9)  2 1/ 3
Complement   1  ↓ 2  3   (Quotient x Complement)
                 2 3 / 6
```

Here, the divisor is nearer to the base 10.

Complement = 10 – 9 = 1

- Since the base has one zero, one digit will be separated by a slash for th remainder column. In the above example, the right-most 3 is separated by a slash.
- Carry down the first digit (2) of the dividend. This is the first digit of the quotient.
- Multiply the first digit of the quotient by the complement and place it in the dividend column; next to the first digit of the dividend. i.e. 2 x 1 = 2
- Write mechanically the sum of the digits of the second column to get the second digit of the quotient. Here 1 + 2 = 3 is the second digit of the quotient.
- Multiply the second digit of the quotient by the complement and write it next to the second digit of the quotient. Place 3 x 1 = 3 in the remainder column.
- Write the sum of the two digits of the remainder column. Since the remainder 6 is less than the divisor, we have completed the division.
- Hence, Quotient = 2 3; Remainder = 6

**Example 3**: Divide 10025 by 88

**Solution**:     Base $= 100$

Complement $= 100 - 8\ 8 = 1\ 2$

| Column 1 | | | Column 2 (Q) | | | Column 3 (R) | |
|---|---|---|---|---|---|---|---|
| Divisor | 8 | 8 | 1 | 0 | 0 | 2 | 5 |
| Complement $=$ | 1 | 2 | | 1 | 2 | – | – |
| | | | | | | 2 | – |
| | | | ▼ | ▼ | ▼ | 3 | 6 |
| | | | 1 | 1 | 3 | 8 | 1 |

Hence,

Quotient $= 113$

Remainder $= 81$

Explanation:

1.  Here Divisor $= 88$

    Nearest base in power of $10 = 100$

    Complement $=$ Base $-$ Divisor

    $= 100-88 = 12$

2.  Arrange the digits in the columns as shown above, separating quotient and remainder. Since there are two zeros in the base, the remainder column will have the two right-most digits of the dividend.

3.  Carry down 1 of column 2 (the first digit of the dividend). This is the first digit of the quotient.

4.  Multiply the first digit of the quotient with the complement and place it in the dividend column, next to the first digit of the dividend. 12 X 1 $= 12$ is placed below 0.

| | Column 1 | | Column 2 (Q) | | | Column 3 (R) | |
|---|---|---|---|---|---|---|---|
| | 8 | 8 | 1 | 0 | 0 | 2 | 5 |
| C = | 1 | 2 | | 1 | 2 | – | – |
| | (Divisor) | | | | | | |
| | | | 1 | 1 | | | |

5. Carry down the sum of the circled digits; this will give you the second digit of the quotient. The second digit of quotient is 0 + 1 = 1

6. Multiply the second digit of the quotient with the complement and place it in the dividend column, next to the second digit of the dividend.    12 x 1 = 12 is placed below the second zero.

| | Column 1 | | Column 2 (Q) | | | Column 3 (R) | |
|---|---|---|---|---|---|---|---|
| | 8 | 8 | 1 | 0 | 0 | 2 | 5 |
| C | 1 | 2 | | 1 | 2 | – | – |
| | | | | | 1 | 2 | |
| | | | 1 | 1 | 3 | | |

7. Carry down the sum of the second circled digits; this will give you the third digit of the quotient. The third digit of the quotient is 0 + 2 + 1 = 3

8. Multiply the complement 12 with the third quotient (3) and write it below the fourth digit of the remainder. 12 x 3 = 36 is placed below 2 in the column 3.

| | Column 1 | | Column 2 (Q) | | | Column 3 (R) | |
|---|---|---|---|---|---|---|---|
| | 8 | 8 | 1 | 0 | 0 | 2 | 5 |
| C = | 1 | 2 | | 1 | 2 | – | – |
| | | | | | 1 | 2 | – |
| | | | | | | 3 | 6 |
| | | | 1 | 1 | 3 | | |

9.  Sum up the digits of Column 3 to get the remainder. The above process repeats until the digit thus obtained in the remainder column is less than the original divisor. Here the sum of the digits of Column 3 = 81 and 81< 88 (Divisor), hence Quotient = 113 and Remainder = 81

**Example 4:** Divide 1121134 by 8988

**Solution:**

| Column 1 | | | | Column 2 (Q) | | | Column 3 (R) | | | |
|---|---|---|---|---|---|---|---|---|---|---|
| Divisor | 8 | 9 | 8 | 8 | 1 | 1 | 2 | 1 | 1 | 3 | 4 |
| Complement = | 1 | 0 | 1 | 2 | | | | – | – | – | – |
| | | | | | 1 | | | | | | |

Explanation:
1.  Here Divisor = 8988
    Nearest base in power of 10 = 10000
    Complement = Base – Divisor
    $$= 10000 - 8988 = 1012$$
2.  Arrange the digits in the columns as shown above, separating quotient and remainder. Since there are four zeros in the base, the remainder column will contain 4 digits.
3.  Carry 1 of column 2 (the first digit of the dividend) down. This is the first digit of the quotient.
4.  Multiply the first digit of the quotient with the complement and place it in the dividend column, next to the first digit of the dividend. 1012 x 1 = 1012 is placed below 1.

|  | Column 1 | Column 2 (Q) | Column 3 (R) |
|---|---|---|---|
| Divisor | 8 9 8 8 | 1 1 2 | 1 1 3 4 |
| Complement = | 1 0 1 2 | 1 0 | 1 2 |
|  |  | 1 2 |  |

5. Add the second digit of column 2; this will give you the second digit of the quotient. The second digit of the quotient is $1 + 1 = 2$

6. Multiply the second digit of the quotient with the complement and place it in the dividend column, next to the second digit of the dividend. $1012 \times 2 = 2024$ is placed below the second zero.

|  | Column 1 | Column 2 (Q) | Column 3 (R) |
|---|---|---|---|
| Divisor | 8 9 8 8 | 1 1 2 | 1 1 3 4 |
| Complement = | 1 0 1 2 | 1 0 | 1 2 |
|  |  | 2 | 0 2 4 |
|  |  | 1 2 4 |  |

7. Carry down the sum of the third digit to the quotient column; this will give you the third digit of the quotient. The third digit of the quotient is $2 + 0 + 2 = 4$

8. Multiply the complement 1012 with the third quotient (4) and write it below 0 in the remainder column as shown below. $1012 \times 4 = 4048$.

| | Column 1 | Column 2 (Q) | Column 3 (R) |
|---|---|---|---|
| Divisor | 8 9 8 8 | 1 1 2 | 1 1 3 4 |
| Complement = | 1 0 1 2 | 1 0 | 1 2 |
| | | 2 | 0 2 4 |
| | | | 4 0 4 8 |
| | | 1 2 4 | 6 6 2 2 |

9. Add the digits of Column 3 to get the remainder. The sum here is 6622, which is less than the divisor, hence the division process is complete.

10. Quotient = 124 and Remainder = 6622

*Case 2: When the remainder is more than the divisor.*

**Example 5**: Divide 10312 by 87?

**Solution**: Base = 100

Complement = 100 – 8 7 = 13

Number of digits to be placed in remainder column = 2

Follow the same method of calculation until you get the **Remainder > Dividend**

| | Column 2 (Q) | Column 3 (R) | |
|---|---|---|---|
| 8 7 | 1 0 3 | 1 2 | |
| C = 1 3 | 1 3 | – – | (13 X 1 = 13) |
| | 1 | 3 | (13 x 1 = 13) |
| | | 9 1 | (13 x 7 = 91) |
| | 1 1 7 | 13 3 | |

Since the remainder 133 > 87, add the complement 13 in the remainder column.

|       |       | Column 2 (Q) | Column 3 (R) |                |
|-------|-------|--------------|--------------|----------------|
|       | 8 7   | 1  0  3      | 1   2        |                |
| C = 1 | 3     |    1  3      | —   —        | (13 X 1 = 13)  |
|       |       |       1      | 3            | (13 x 1 = 13)  |
|       |       |              | 9   1        | (13 x 7 = 91)  |
|       |       | 1  1  7      | 1  3 3       |                |
|       |       |              | +  1 3       |                |
|       |       | 1  1  7      | 1  4 6       |                |
|       |       | 1  1  8      |    4 6       |                |

Since the base is 100, the number in the remainder column should be less than 100; therefore the left-most 1 of the remainder column will be transferred to the quotient column. Thus Quotient = 117 and Remainder = 46.

## The importance of zero in the complement of divisor

While taking the complement from the base, we need to be very careful. If the base is 100, the complement should have two digits and if the base is taken as 1000, the complement will have three digits. Suppose the divisor is 9, then its complement from the base will be written as 1. For 99 and 999 as divisor, the complement from their respective bases 100 and 1000, will be written as 01 and 001. Let us take an example to understand the modus operandi.

**Example 6:** Divide 11199171 by 99979

**Solution:** Base = 100000
Complement = 00021
Number of digits to be placed in Remainder column = 5

|  | Column 2 (Q) | Column 3 (R) |
|---|---|---|
| 9 9 9 7 9 | 1 1 1 | 9 9 1 7 1 |
| 0 0 0 2 1 | 0 0 | 0 2 1 – – |
|  | 0 | 0 0 2 1 – |
|  |  | 0 0 0 2 1 |
|  | 1 1 1 | 1 0 1 5 0 2 |
|  |  | + 0 0 0 2 1 |
|  | 1 1 1 | 1 0 1 5 2 3 |

Hence, Quotient = 112
Remainder = 0 1 5 2 3

## Paravartya Yojyet (परावर्तयोजयेत)

The literal meaning of this sutra is **Transpose and Apply**. There is a very small difference between the Nikhilam and Paravartya Yojyet methods. This sutra works effectively when the first digit of the divisor is 1. First, find the complement from the base as done in Nikhilam. This complement will be revised by Paravartya sutra by writing each digit of the complement with a changed sign separately. For example, if the divisor is 87, the nearest base = 100 and complement = 100 – 87 = 13 and revised complement = – 1 – 3. This revised complement will now be the basis of division.

**Example 7**: Divide 1358 by 113

**Solution:**     Base = 100
        Complement = 113 – 100 = 13
        Complement with changed sign for each digit = – 1 – 3

| | Column 1 | | | Column 2(Q) | | Column 3(R) | |
|---|---|---|---|---|---|---|---|
| | 1 | 1 | 3 | 1 | 3 | 5 | 8 |
| Complement = | | 1 | 3 | | −1 | −3 | − |
| Revised Complement= | | −1 | −3 | | | −2 | −6 |
| | | | | 1 | 2 | 0 | 2 |

Explanation:−

1.  Base $=100$

    Complement $= 113 - 100 = 13$

    Writing each digits of complement with the changed sign
    $= -1- 3$.

    Since the number of zeros in the base is 2, the number of digits in the remainder column is 2.

2.  Carry 1 **(the fist digit of Column 2)** down; this will give you the first digit of the quotient.

| | Column 1 | | | Column 2(Q) | | Column 3(R) | |
|---|---|---|---|---|---|---|---|
| | 1 | 1 | 3 | 1 | 3 | 5 | 8 |
| Complement = | | 1 | 3 | | | | |
| Revised Complement= | | −1 | −3 | | | | |
| | | | | 1 | | | |

3.  Multiply the first digit of the quotient with the **revised complement** and write it next to the first digit of the dividend.

    $-1 - 3 \times 1 = -1 - 3$

    Now, add each digit of Column 2 and Column 3 to get the Quotient and Remainder. Here, the sum of the digits of the second column is $3 - 1 = 2$. Now multiply the revised complement with the second digit of the quotient and write it next to the second digit of the quotient.

    $-1 - 3 \times 2 = -2 - 6$

    Add the digits of column 3 to get the remainder.

| | Column 1 | Column 2(Q) | | Column 3(R) | |
|---|---|---|---|---|---|
| | 1 1 3 | 1 | 3 | 5 | 8 |
| Complement = | 1 3 | | −1 | −3 | |
| Revised Complement= | −1 −3 | + | | −2 | −6 |
| | | 1 | 2 | 0 | 2 |

**Example 8: Divide 239479 by 11213**

**Solution:** Base = 10000

Complement = 11213 − 10000 = 1213

Complement with changed sign for each digit = − 1 − 2 − 1 − 3

Since the base has 4 zeros, arrange the digits in the quotient and remainder columns accordingly.

| | Column 1 | Column 2 (Q) | Column 3 (R) | |
|---|---|---|---|---|
| Divisor | 1 1 2 1 3 | 2  3 | 9  4  7  9 | |
| Complement = 1 2 1 3 | | −2 | −4 −2 −6 −0 | (Revised complement x 2) |
| Revised Complement= −1 −2 −1 −3 | | | −1 −2 −1 −3 | (Revised complement x 1) |
| | | 2  1 | 4  0  0  6 | |

Quatient = 21

Remainder = 4 0 0 6

## Urdhva – Tiryagbhyam (उर्ध्व तिर्यगभ्याम्)

The division method studied so far has limited application and is useful when the divisor is near the base. The division method we are going to study now is a mix of the Vedic sutra, Urdhva Tiryagbhyam and sub-sutra, Dhwajanka. This is also known as straight division because we get the answer as a quotient and remainder in one line. Thus we save time and space. In straight division; divisor is bifurcated as operator and Dhwajank so that we go on dividing not by the whole divisor, but by the small operator and go on modifying the sub dividend portions using the dhwajank and quotient by the Urdhva Tiryak sutra.

**Rules:**

- Irrespective of the number of digits in the divisor, our divisor is of one digit or two digits, depending upon the situation. The remaining digits of the divisor are placed at the top of that number. For example – If we are to divide x y z p q r by a b c, then we shall write

$$a^{bc} | \quad x \quad y \quad z \quad p \quad | \quad q \quad r$$

  Quotient / Remainder

  Here our new divisor is a and bc is called the flag (Dhwajank) digit.

- The number of digits on the right side of the slash will depend on the number of flag digits on the top.

- Divide the first digit of the dividend by the divisor and write the quotient in the quotient column and the remainder is placed before the next dividend digit to give gross dividend.

- Subtract the product of the flagged digit and the first quotient digit from the gross dividend to get the net dividend.

- Now this net dividend is to be divided by the divisor. The net dividend should always be positive. In case you obtain a negative net dividend, make an adjustment by reducing the quotient and recalculating the remainder thereafter.

- The process of getting gross dividend and net dividend is to be repeated till you reach the desired result.

- The main problematic area in this straight division is subtraction from the gross dividend the product of the quotient and flagged digits. The steps written here will help you in this regard.

- When there is a single flagged digit, we subtract the product of the last quotient digit and the flagged digit at each step.
- When there are two flag digits, in the first step product of the first flagged digit and the first quotient digit is subtracted and from the second step onwards, the cross product of the two flagged digits and the last two quotient digits is subtracted.
- When there are three flagged digits: – first subtract the product of the first flagged digit and the first quotient digit. In the second stage, the cross product of the first two flagged digits and first two quotient digits will be subtracted from the gross dividend. In the third stage, the cross product of the three flagged digits and the three quotients will be subtracted. After the third stage, the subtraction of the cross product of the three flagged digits and the three quotients will be continued.
- Let us take some examples to understand the modus operandi more clearly.

**Example 9:** Divide 1764 by 42

**Solution:**

Step 1: Take 4 as the main divisor and 2 as the flag digit. Arrange the divisor, flag digit, quotient and digit for remainder, if any, according to the rule explained above. Since we have taken only one digit as the flagged digit, the unit digit will be put in the remainder column.

| $4^2$ | 1 | 7 | 6 | 4 |
|---|---|---|---|---|
| | | Quotient | | Remainder |

Step 2: Divide 17 by 4

First digit of quotient = 4

Remainder = 1

Put the quotient 4 in quotient column and the remainder 1 will be place before 6 in the dividend column.

$$\begin{array}{c|ccc} 4^2 & 1 & 7 & {}_16 & 4 \\ \hline & & 42 & & \end{array}$$

Gross Dividend = 16

Net Dividend = Gross Dividend – First digit of quotient x Flag digit

= 16 – 2 x 4 = 8

Step 3: Divide 8 by 2

Quotient = 2

Remainder = 0

$$\begin{array}{c|ccc} 4^2 & 1 & 7 & {}_16 & {}_04 \\ \hline & & & & \end{array}$$

Now our calculation has moved in remainder part. Here Gross Dividend = 4

Net Dividend = 4 – 2 x 2 = 0

Since, we are in the remainder part of division, our division process is completed.

Hence 1764 ÷ 42, Q = 42 R = 0

**Example 10:** Divide 387 by 32

**Solution:**

Step 1: Here, we have a double-digit divisor, so 3 is taken as the main divisor for the whole operation and 2 is the flag digit (Dhwajank). Since the flag digit is a single digit, so is the remainder. Hence the remainder digit is 1

$$\begin{array}{c|cc} 3^2 & 3 & 8 & 7 \\ \hline & & & \end{array}$$

Step 2: Divide 3 by 3. Quotient = 1 and Remainder = 0
Put the quotient in thequotient column and remainder 0
before 8 in the dividend column.

$$\begin{array}{c|ccc} 3^2 & 3 & _08 & 7 \\ \hline & & 1 & \end{array}$$

Gross Dividend = 08
Net Dividend = 08 – first digit of quotient x flag digit
= 08 – 1 x 2 = 6

Step 3: Divide 6 by 3
Quotient = 2 and Remainder = 0.
0 is now placed before 7 in the dividend column, making
it 07, the next dividend.

$$\begin{array}{c|ccc} 3^2 & 3 & _08 & _07 \\ \hline & & 1\ 2 & \end{array}$$

Gross Dividend = 07
Net Dividend = 07 – second digit of quotient x flag digit
= 07 – 2 x 2 = 3
Since,our calculation has moved inare in remainder side,
we do not do the division.

$$\begin{array}{c|ccc} 3^2 & 3 & _08 & _07 \\ \hline & & 1\ 2 & 3 \end{array}$$

Hence, 387 ÷32, Quotient = 12, Remainder = 3

**Example** 11: Divide 38982 by 73

**Solution**:

The first step in the Dhvajanka method is to make the column
as shown below.

Step 1: Out of the divisor 73, we put down only the first digit
i.e. 7 in the divisor- column and put the other digit i.e. 3 on
the top of the flag.

Here main divisor $=7$
Dhvajanka $=3$ (one digit)

$$7^3 \mid \quad 3 \quad 8 \quad 9 \quad 8 \mid 2$$

Hence, a line is drawn vertically from the right, leaving one digit at the end in the remainder column as shown above. Now the entire division is to be carried out by 7.

Step 2: As the first digit from the left of dividend (3) is less than 7, we take 38 as our first dividend. Divide 38 by 7

$$Q_1 = 5 \quad R_1 = 3$$

$Q_1$ = The first quotient
$R_1$ = The first remainder.
Put the quotient below the horizontal line and prefix the remainder 3 below the digit 9.

$$7^3 \mid \quad 3 \quad 8 \quad {}_3 9 \quad 8 \mid 2$$
$$\underline{\phantom{7^3 \mid \quad 3 \quad 8}}$$
$$5$$

Step 3:

New gross dividend = 39
Subtract the product of dhvajanka 3 and the first quotient (5) i.e. 3 x 5 = 15 from 39.
Net divided = 39 – 3 x 5 = 24
Divide 24 by 7

$$Q_2 = 3 \quad R_2 = 3$$

$Q_2$ = The second quotient
$R_2$ = The second Remainder
Prefix the remainder 3 below 8 above the horizontal line

$$7^3 \mid \quad 3 \quad 8 \quad {}_3 9 \quad {}_3 8 \mid 2$$

Step 4:

New gross dividend =38
Net Dividend = Gross Dividend – product of $2^{nd}$ quotient and dhvajanka
= 38 – 3 x 3 =29

Net dividend = 29

Divide it by 7

$$Q_3 = 4 \quad R_3 = 1$$

$Q_3$ = The third digit of quotient

$R_3$ = The third digit of remainder

Place 4 and 1 as discussed above.

$$7^3 \ \overline{\left| \begin{array}{ccccc} 3 & 8 & {}_3 9 & {}_3 8 & {}_1 2 \end{array} \right.}$$
$$\qquad 5 \quad 3 \quad 4$$

Step 5:

New Gross dividend = 12

$$7^3 \ \overline{\left| \begin{array}{ccccc} 3 & 8 & {}_3 9 & {}_3 8 & {}_1 2 \end{array} \right.}$$
$$\qquad 5 \quad 3 \quad 4 \qquad 0$$

Net Dividend = Gross Dividend - product of dhvajanka and third quotient

$$= 12 - 3 \times 4 = 0$$

Since, we are in remainder part so we stop division process.

Hence, **Quotient = 534** and **Remainder = 0**

**Example 12:** Divide 72 38 761 by 524?

**Solution:** If we take two digits as dhvajanka, then the remainder column will certainly have two digits. See the arrangement below.

$$5^{24} \ \overline{\left| \begin{array}{ccccc|cc} 7 & 2 & 3 & 8 & 7 & 6 & 1 \end{array} \right.}$$

Step 1: Divide 7 by 5

**$Q_1$ = 1 and $R_1$ = 2**

$$5^{24} \ \overline{\left| \begin{array}{ccccc|cc} 7 & {}_2 2 & 3 & 8 & 7 & 6 & 1 \end{array} \right.}$$
$$\qquad 1$$

Step 2: Gross dividend = 22

Net dividend = Gross Dividend – Product of the first quotient and the first flag digit

$$= 22 - 1 \times 2 = 20$$

Step 3: Divide 20 by 5

$Q_2 = 4$ and $R_2 = 0$

Since $R_2 = 0$, we have to take the quotient below 4 i.e. New quotient $= 4 - 1 = 3$. This is because we cannot consider the remainder zero in the middle of the division. The same process will be applicable in case the net divisor in the middle is either zero or negative.

Hence, **Revised Quotient ( $Q_2$ ) = 3 and $R_2$ = 5**

$$\begin{array}{c|ccccccc} 5^{24} & 7 & {}_2 2 & {}_5 3 & 8 & 7 & 6 & 1 \\ \hline & & 1 & 3 \end{array}$$

Step 4: Gross dividend $= 53$

Net dividend $= 53 -$ (sum of cross product of two flag digits and two quotients)

$$= 53 - (2 \times 3 + 1 \times 4)$$
$$= 43$$

Divide 43 by 5

$$Q_3 = 8 \text{ and } R_3 = 3$$

$$\begin{array}{c|ccccccc} 5^{24} & 7 & {}_2 2 & {}_5 3 & {}_3 8 & 7 & 6 & 1 \\ \hline & & 1 & 3 & 8 \end{array}$$

Step 5: Gross dividend $= 38$

Net dividend $= 38 -$ (sum of cross product of 38 and 24)

$$= 38 - (3 \times 4 + 2 \times 8)$$
$$= 38 - 28 = 10$$

Divide 10 by 5

$Q_4 = 2$ and $R_4 = 0$

Since $R_4 = 0$ so we have to take $Q_4 = 1$(discussed above)

For $Q_4 = 1$ and $R_4 = 5$

$$\begin{array}{c|ccccccc} 5^{24} & 7 & {}_2 2 & {}_5 3 & {}_3 8 & {}_5 7 & 6 & 1 \\ \hline & & 1 & 3 & 8 & 1 \end{array}$$

Step 6: Gross dividend = 57

Net dividend = 57 – (sum of cross product of 24 and 81)

$\qquad$ = 57 – (2 x 1 + 4 x 8)

$\qquad$ = 57–34

$\qquad$ = 23

Divide 23 by 5

$\qquad$ $Q_5$ = 4 and $R_5$ = 3

Remainder = 361 – (sum of cross product of 24 and 14) x 10 – (last flag digit x last quotient)

$\qquad$ = 361 – (8 +4) x 10 – (4 x 4)

$\qquad$ = 361–120-16= 225

$$5^{24} \mid 7 \quad _2 2 \quad _5 3 \quad _3 8 \quad _5 7 \mid _3 6 \quad 1$$
$$\overline{\phantom{5^{24} \mid} 1\ 3\ 8\ 1\ 4 \qquad\qquad 225}$$

Hence Quotient = 13814 and R = 225

**Example 13**: Divide 9862145 by 20132

**Solution**: Arrange the dividend, divisor and remainder digit as discussed in the above examples.

$$20^{132} \mid 9 \quad 8 \quad 6 \quad 2 \mid 1 \quad 4 \quad 5$$

Step 1: For convenience, we have taken the dividend of two digits and flag digit of three digits.

$\qquad$ Divide 98 by 20

$\qquad$ Q = 4 , R = 18

$\qquad$ Put the quotient below the horizontal line and prefix the remainder 18 below the digit 6.

$$20^{132} \mid 9 \quad 8 \quad _{18}6 \quad 2 \mid 1 \quad 4 \quad 5$$
$$\overline{\phantom{20^{132} \mid}\ 4}$$

$\qquad$ Gross Dividend = 186

$\qquad$ Net Dividend = 186 – 1 x 4

$\qquad$ = 182

Step 2: Divide 182 by 20
   Q = 8 , R = 22
Put the quotient below the horizontal line and prefix the remainder 22 below the digit 2.

$$20^{132} \,\big|\, 9 \quad 8 \quad {}_{18}6 \quad {}_{22}2 \,\big|\, 1 \quad 4 \quad 5$$

$$\big|\; 4 \quad 8$$

Gross Dividend = 222
Net Dividend = 222 – Cross product of 13 and 48
      = 202

Step 3: Divide 202 by 20
   Q = 9 , R = 22
Put the quotient below the horizontal line and prefix the remainder 22 below the digit 1 in the remainder column.

$$20^{132} \,\big|\, 9 \quad 8 \quad {}_{18}6 \quad {}_{22}2 \,\big|\, 1 \quad 4 \quad 5$$

$$\big|\; 4 \quad 8 \quad 9$$

Gross Dividend = 221
Net Dividend = 221 – Cross product of 132 and 489
      = 221 – 41   1   3   2

Step 4: Divide 180 by 20
   Q = 8 , R = 20
   Put the quotient below the horizontal line and prefix the remainder 20 below the digit 4 in the remainder column.

$$20^{132} \,\big|\, 9 \quad 8 \quad {}_{18}6 \quad {}_{22}2 \,\big|\, {}_{22}1 \quad {}_{20}4 \quad 5$$

$$\big|\; 4 \quad 8 \quad 9 \qquad 8$$

Gross Dividend = 204
Net Dividend = 204 – Cross product of 132 and 898
      = 204 – 51   8   9   8
      = 153

Step 5: Divide 153 by 20
   Q = 7 , R = 13
Put the quotient below the horizontal line and prefix the remainder 13 below the digit 5 in the remainder column.

$$20^{132} \mid 9 \quad 8 \quad {}_{18}6 \quad {}_{22}2 \mid {}_{22}1 \quad {}_{20}4 \quad {}_{13}5$$
$$\phantom{20^{132} \mid} \; 4 \; 8 \quad 9 \qquad \; 8 \quad 7$$

Since, we are left with two digits in the remainder column, we can stop working.

Hence, 9862145 ÷ 20132 = 489.87

**Example 14:** Divide 5362968527 by 9213649875

**Solution:** At first look, this sum seems horrible as the number of digits in the divisor is 10. The conventional method will take much of your time in the hit and trial method of multiplication, but the Vedic method of division makes this sum child's play. For convenience, we will take the actual divisor of one digit and the dhwajank of the rest of the 9 digits. As discussed earlier, the number of digits with the flag decides the number of digits to be kept in the remainder.

$$9^{213649875} \mid 5 \quad 3 \quad 6 \quad 2 \quad 9 \quad 6 \quad 8 \quad 5 \quad 2 \quad 7$$

Step 1: Here, 5 < 9, so we put a decimal at the beginning.

Now divide 53 by 9

Q = 5, R = 8

Gross Dividend = 86

Net Dividend = 86 – first flagged digit x quotient

= 86 – 5 x 2

= 76

$$9^{213649875} \mid 5 \mid 3 \quad {}_{8}6 \quad 2 \quad 9 \quad 6 \quad 8 \quad 5 \quad 2 \quad 7$$
$$\phantom{9^{213649875} \mid 5 \mid} 0.5$$

Step 2: Divide 76 by 9

Q = 8 , R = 4

Gross Dividend = 42

Net Dividend = 42 – cross product of 21 and 58

$= 42 - 21$

$= 21$

| $9^{213649875}$ | 5 | 3 | $_8$6 | $_4$2 | 9 | 6 | 8 | 5 | 2 | 7 |
|---|---|---|---|---|---|---|---|---|---|---|
| | 0.5 | 8 | | | | | | | | |

Step 3: Divide 21 by 9

$Q = 2$, $R = 3$

Gross Dividend $= 39$

Net Dividend $= 39$ − cross product of 213 and 582

$= 39 - 27$

$= 12$

| $9^{213649875}$ | 5 | 3 | $_8$6 | $_4$2 | $_3$9 | 6 | 8 | 5 | 2 | 7 |
|---|---|---|---|---|---|---|---|---|---|---|
| | 0.5 | 8 | 2 | | | | | | | |

Step 4: Divide 12 by 9

$Q = 1$, $R = 3$

| $9^{213649875}$ | 5 | 3 | $_8$6 | $_4$2 | $_3$9 | $_3$6 | 8 | 5 | 2 | 7 |
|---|---|---|---|---|---|---|---|---|---|---|
| | 0.5 | 8 | 2 | 1 | | | | | | |

Since we have reached upto 4 decimal places, there is no need to move any further.

**7**

# Square

## Introduction

Squaring a number means multiplication of the number by itself. Mathematically, a x a = $a^2$. Here is the geometrical representation of first five square numbers.

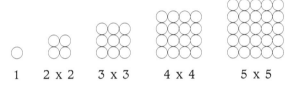

| 1 | 2 x 2 | 3 x 3 | 4 x 4 | 5 x 5 |

The school curriculum follows only two methods for finding the square of a number.

- By multiplying the number by itself through long multiplication process.

  Example: $(12)^2$ = 12 x 12 = 144

- By Algebraic Expansion:

  Here, we generally use either of the two formulae—

    i) $(a+b)^2 = a^2 + 2ab + b^2$

    ii) $(a- b)^2 = a^2 - 2ab + b^2$

Example (a): $(13)^2 = (10 + 3)^2$

= $10^2$ + 2 x 10 x 3 + $3^2$

= 100 + 60 + 9 = 169

Example (b): $(96)^2 = (100 - 4)^2$

= $100^2$ - 2 x 100 x 4 + $4^2$

= 10000 - 800 + 16

= 9216

But in Vedic mathematics, there are around 5 methods to find the square of a number. Moreover, the Vedic method is 10 times faster than the traditional method. Some of these methods have limited application but there is a special method called the Duplex method which is said to be the gem of all methods.

# Vedic Method

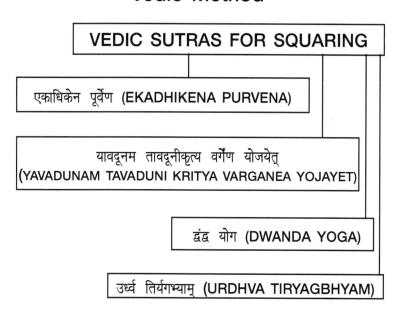

**VEDIC SUTRAS FOR SQUARING**

एकाधिकेन पूर्वेण (EKADHIKENA PURVENA)

यावदूनम तावदूनीकृत्य वर्गेण योजयेत्
(YAVADUNAM TAVADUNI KRITYA VARGANEA YOJAYET)

द्वंद्व योग (DWANDA YOGA)

उर्ध्व तिर्यगभ्याम् (URDHVA TIRYAGBHYAM)

## Meaning of Vedic Sutra

1.  **Ekadhikena Purvena (एकाधिकेन पूर्वेण):** This Vedic method of finding a square has limited application and is valid only when the unit digit of any number is 5. The answer comes into two parts. The RHS part of the answer is the square of 5 and LHS part of the answer is the product of the remaining digit and its Ekadhikena. In case the number before 5 is bigger, the multiplication of the number and its Ekadhikena can be carried out by the long method of multiplication.

2. **Yavadunam Tavaduni kritya vargena yojayet** (यावदूनम तावदूनी कृत्य वर्गेण योजयेत): This Vedic sub-sutra is used for squaring numbers which are closer to the base $(10^n)$. With a little practice, though, you can extend it to numbers which are farther from the base using the sub-base provided the sub-base is a multiple of $10^n$.

   This Vedic-sutra simply says –

   (a) Find the extent or deficiency of a number to be squared with respect to its base. This extent or deficiency is termed here as the **deviation**.

   (b) Set up the square of the deviation at the end.

3. **Dwanda-Yoga or Duplex Mehod** (द्वंद्व योग): The two Vedic methods discussed above have limited application. So the question is – what will you do if the number that is to be squared does not satisfy either of the above condition?

   The Duplex combination is applicable in all the cases. The term Dwanda-Yoga is used in two different senses– The first one is by squaring and the second one is by cross multiplication.

   i)  In the case of a single central digit **a**, the duplex is its square. i.e. $a^2$

   ii) In the case of an even number of digits (say a and b) equidistant from the two ends, the duplex is taken as double the cross product of a and b (i.e.2ab)

4. **Urdhva–Tiryak** (उर्ध्व तिर्यक): Squaring a number means multiplying a number twice by itself. This method of squaring is nothing new but it is the same as that of the multiplication of two numbers in the Cross and Dot method discussed in the chapter on Multiplication. It is therefore expected from readers to do the squaring of any number by Urdhva-Tiryak Method, which is nothing but the multiplication of a single number twice.

## Ekadhikena Purvena (एकाधिकेन पूर्वेण)

As discussed above, this Vedic sutra is applicable when the number to be squared has 5 as its unit digit. Mathematically,

$$(A\ 5)^2 = A \times (A+1)\ /\ 5^2$$

Here, the Ekadhikena of A is A+1. Let us understand the modus operandi of this method as discussed above, with the help of some examples.

**Example 1:** Find the square of 85

**Solution:** At the beginning, I said that the answer is split into two parts.

RHS = Square of 5 = 25
LHS = Remaining digit x Next digit
  = 8 x 9 = 72

Hence, $(85)^2$ = 7225

**Example 2:** Find the square of 55

**Solution:**

RHS = Square of 5 = 25
LHS = Remaining digit x Next digit
    = 5 x 6 = 30

Hence, $(55)^2$ = 3025

**Example 3:** Find the square of 125

Solution:

RHS = Square of 5 = 25
LHS = Remaining digit x Next digit
    = 12 x13 = 156

Multiplication of 12 x 13 can be done by the Nikhilam method.

12 + 2
13 + 3
15 / 6

Moreover, multiplication in such a case can be done by the Urdhva-Tiryak method.

    1 2
x 1 3

$$1 \times 1 / 1 \times 3 + 1 \times 2 / 2 \times 3$$
$$= 1\ 5\ 6$$
Hence, $(125)^2 = 15625$

**Example** 4: Find the square of 165

Solution:

RHS = Square of 5 = 25
LHS = Remaining digit x Next digit
    = 16 x17 = 156

Multiplication of 12 x 13 can be done by the Nikhilam method.

16 + 6
17 + 7
23 / 42

$$= 272$$

Moreover, multiplication in such case can be done by the Urdhva-Tiryak method.

       1 6
     x 1 7
$$1 \times 1 / 1 \times 6 + 1 \times 7 / 6 \times 7$$
$$= 1 / 13 / 42$$
$$= 2\ 7\ 2$$

Hence, $(165)^2 = 27225$

**Example** 5: Find the square of 245

**Solution:**

RHS = Square of 5 = 25
LHS = Remaining digit x Next digit
    = 24 x 25 = 600

(See special method of multiplication by 25 in *Multiplication through Observation*.)
Hence, $(245)^2 = 60025$

## Yavadunam Tavaduni kritya vargena yojayet
(यावदूनम तावदूनी कृत्य वर्गेण योजयेत)

This sutra works better when the number to be squared is near the base 10, 100, 1000 ... or is the multiple of the base i.e. 20, 30, 40, --- 200, 300, 400--- etc. Let us divide the squaring concept through this Vedic sub-sutra in two parts.

*Case 1: When the number is near the base 10, 100, 1000...... $10^n$*

The answer is arrived at in two parts.

LHS = Number + Deviation

(Deviation may be positive or negative, depending on the base)

RHS = Square of deviation

The RHS will contain the same number of digits as the number of zeros in the base. The excess digit if any will be carried over to LHS and the deficit digit, if any, will be filled up by putting the zeros to the left of the RHS.

**Example 6:** Find the square of 13

**Solution:** Number 13 is closer to base 10.

$$\text{Deviation} = 13 - 10 = 3.$$
$$(13)^2 = 13 + 3 / 3^2$$
$$= 169$$

**Example 7:** Find the square of 16

**Solution:** Number 16 is closer to base 10.

$$\text{Deviation} = 16 - 10 = 6.$$
$$(16)^2 = 16 + 6 / 6^2$$
$$= 22 / 36$$
$$= 256$$

Since Base = 10, the RHS will contain single digit.

**Example 8:** Find the square of 91?

**Solution:** Number 91 is closer to base 100.

$$\text{Deviation} = 91 - 100 = -9$$
$$(91)^2 = 91 - 9 \ / \ (-9)^2$$
$$= 82/ \ 81$$

**Example 9:** Find the square of 97

**Solution:** Number 97 is closer to base 100.

$$\text{Deviation} = 97 - 100 = -3$$
$$(97)^2 = 97 - 3 \ / \ (-3)^2$$
$$= 94/ \ 9$$

Since the Base $=100$, the RHS should have 2 digits, so one additional zero will be placed before 9.

$$(97)^2 = 9409$$

*Case 2: When the base is not in the form of $10^n$, but the multiple of 10.*

If the number to be squared is near the base 20, 30, 40, ---- or 200, 300, 400, ---- or 2000, 3000, 4000, --- the Yavadunam Tavduni sub-sutra will work with a slight change.

The answer will be arrived at in two parts.

The RHS part of the answer will be the square of the deviation from the base. The LHS part of the answer should be written with utmost care. LHS = (Number to be squared + Deviation) x sub-base.

**Example 10:** Find the square of 32

**Solution:** Number 32 is closer to base 30.

$$\text{Deviation} = 32 - 30 = 2$$
$$30 = 3 \text{ x } 10$$
$$\text{Sub-base} = 3$$
$$\text{Actual base} = 10$$

$(32)^2$ = $(32 + 2) \times 3 / (2)^2$
       = $102/ 4$
       = $1024$

**Example** 11: Find the square of 47

**Solution:** Number 47 is closer to base 50.

Deviation = $47 - 50 = -3$
        $50 = 5 \times 10$
        Sub-base = 5
        Actual base = 10
$(47)^2$ = $(47 - 3) \times 5 / (-3)^2$
       = $220/ 9$
       = $2209$

**Example** 12: Find the square of 204

**Solution:** 204 is closer to base 200.

Deviation = $204 - 200 = 4$
        $200 = 2 \times 100$
        Sub-base = 2
        Actual base = 100
$(204)^2$ = $(204 + 4) \times 2 / (4)^2$
        = $416/ 16$
        = $41616$

**Example** 13: Find the square of 482

**Solution:** 482 is closer to base 500.

Deviation = $482 - 500 = -18$
        $500 = 5 \times 100$
        Sub-base = 5
        Actual base = 100
$(482)^2$ = $(482 - 18) \times 5 / (-18)^2$
        = $5 \times 464/ 324$
        = $232 \ 0/3 \ 24$
        = $232324$

Let us take the base 500 and find the square of 482 in another way.

$$500 = 1000/2$$

Hence, Base = 1000 and sub-base = ½

Deviation = 482 −500 = −18

$(482)^2 = (482 −18)$ x ½ / $(−18)^2$

$= 232 / 324$

$= 232324$

**Example 14:** Find the square of 709

**Solution:** 709 is closer to base 700.

Deviation = 709 − 700 = 9

700 = 7 x 100

Sub-base = 7

Actual base =100

$(709)^2 = (709 + 9)$ x 7 / $(9)^2$

$= 7$ x 718/ 81

$= 502681$

**Example 15:** Find the square of 8989

**Solution:** 8989 is closer to base 9000.

Deviation = 8989 − 9000 = −11

9000 = 9 x 1000

Sub-base = 9

Actual base =1000

$(8989)^2 = (8989 −11)$ x 9 / $(−11)^2$

$= 9$ x 8978/ 121

$= 80802 /121$

$= 80802121$

## Duplex or Dwanda yoga (द्वंद्व योग)

The Dwandwa-Yoga or Duplex method of squaring is one of the best squaring methods in Vedic Mathematics. By using this sutra, we can find the square of any number, of any length, with comfort and ease in one line. After a little practice, you can find the square of any number mentally. This is unique in the sense that it has universal application. Let us denote the duplex of a number by **D**.

- Duplex of 1 digit number = Square of that number
  
  $$D(a) = a^2$$
  
  Duplex of 2 = $2^2$ = 4  Duplex of 6 = $6^2$ = 36

- Duplex of 2 digit number = 2 X ( Product of digits)
  
  $$D(ab) = 2ab$$
  
  Duplex of 24 = 2 x ( 2 x 4) = 16
  
  Duplex of 76 = 2 x ( 7 x 6) = 84

- Duplex of 3 digit number = 2 x ( $1^{st}$ digit x $3^{rd}$ digit) + (square of middle digit)
  
  $$D(abc) = 2ac + b^2$$
  
  Duplex of 126 = 2 x (1 x 6 ) + $2^2$ = 16
  
  Duplex of 478 = 2 x (4 x 8 ) + $7^2$ = 113

- Duplex of 4 digit number = 2 x ( $1^{st}$ digit x $4^{th}$ digit) + 2 x ( $2^{nd}$ digit x $3^{rd}$ digit ).
  
  $$D(abcd) = 2ad + 2bc$$
  
  Duplex of 2468 = 2 x ( 2 x 8 ) + 2 x ( 4 x 6 ) = 80
  
  Duplex of 4567 = 2 x ( 4 x 7 ) + 2 x ( 5 x 6 ) = 116

- Duplex of 5 digit number = 2 x ($1^{st}$ digit x $5^{th}$ digit) + 2 x ( $2^{nd}$ digit x $4^{th}$ digit) + (middle digit)$^2$
  
  $$D(abcde) = 2ae + 2bd + c^2$$
  
  Duplex of 16289 = 2 x (1 x 9) + 2 x (6 x 8) + $2^2$ = 118
  
  Duplex of 50406 = 2 x (5 x 6) + 2x (0 x 0) + $4^2$ = 76

- Duplex of 6 digit number = 2 x ($1^{st}$ digit x $6^{th}$ digit) + 2 x ($2^{nd}$ digit x $5^{th}$ digit) + 2 x ($3^{rd}$ digit x $4^{th}$ digit)
  
  $$D(abcdef) = 2af + 2be + 2cd$$

Duplex of 320416 = 2 x (3 x 6 ) + 2 x (2 x 1) + 2 x (0 x 4) = 40

Duplex of 125673 = 2 x (1 x 3 ) + 2 x (2 x 7) + 2 x (5 x 6) = 94

- Duplex of 7 digit number = 2 x ($1^{st}$ x $7^{th}$ digit) + 2 x ($2^{nd}$ x $6^{th}$ digit) + 2 x ($3^{rd}$ x $5^{th}$ digit) + ($4^{th}$ digit)$^2$

    **D (abcdefg) = 2ag +2bf + 2ce + d$^2$**

    Duplex of 2356214 = 2 x (2 x 4) + 2 x (3 x 1) + 2 x (5 x 2) + $6^2$ =78

    Duplex of 1025962 = 2 x (1 x 2) + 2 x (0 x 6) + 2 x (2 x 9) + $5^2$ =65

Once you learn to find the duplex of a number, you need to write the number in groups. The following pattern will help you in grouping the numbers.

$(11)^2 = 1\ 2\ 1$
$(111)^2 = 1\ 2\ 3\ 2\ 1$
$(1111)^2 = 1\ 2\ 3\ 4\ 3\ 2\ 1$
$(11111)^2 = 1\ 2\ 3\ 4\ 5\ 4\ 3\ 2\ 1$
$(111111)^2 = 1\ 2\ 3\ 4\ 5\ 6\ 5\ 4\ 3\ 2\ 1$
$(1111111)^2 = 1\ 2\ 3\ 4\ 5\ 6\ 7\ 6\ 5\ 4\ 3\ 2\ 1$

-------------------------------------------------------------------------------

## Grouping of a number

The grouping of $(24)^2$ will follow the pattern of 1 2 1 of $(11)^2$
The groups for **24** are–

| D(2) | D(24) | D(4) |
|---|---|---|
| 1 digit | 2 digit | 1 digit |

The grouping of $(245)^2$ will follow the pattern of 1 2 3 2 1 of $(111)^{2.}$
The groups of numbers for **245** are-

| D(2) | D(24) | D(245) | D(45) | D(5) |
|---|---|---|---|---|
| 1 digit | 2 digit | 3 digit | 2 digit | 1 digit |

The grouping of $(2456)^2$ will follow the pattern of 1 2 34 3 2 1 of $(1111)^{2.}$

The groups of numbers for **2456** are–

| D(2) | D(24) | D(245) | D(2456) | D(456) | D(56) | D(6) |
|---|---|---|---|---|---|---|
| 1 digit | 2 digit | 3 digit | 4 digit | 3 digit | 2 digit | 1 digit |

| | Summary of the Duplex Method |
|---|---|
| 1 | $D(a) = a^2$ |
| 2 | $D(ab) = 2ab$ |
| 3 | $D(abc) = 2ac + b^2$ |
| 4 | $D(abcd) = 2ad + 2bc$ |
| 5 | $D(abcde) = 2ae + 2bd + c^2$ |
| 6 | $D(abcdef) = 2af + 2be + 2cd$ |
| 7 | $D(abcdefg) = 2ag + 2bf + 2ce + d^2$ |

**How does the Duplex Method work?**

* Form the groups of numbers to be squared as shown above
* Write the duplex of each group
* Once the duplex value for each group is written, add the figures from right to left, keeping only one digit in each separator

**Example** 16: Find the square of 32

**Solution:** The groups for 32 are

$$= \quad \underbrace{3^2}_{D(3)} \quad | \underbrace{2 \times 3 \times 2}_{D(32)} | \quad \underbrace{2^2}_{D(2)}$$

$$= 9 \,|\, 1 \, 2 \,|\, 4$$

$$= 1024$$

**Example** 17: Find the square of 49

**Solution**: The groups for **49** are

$$= \quad \overbrace{4^2}^{D(4)} \quad | \overbrace{2 \times 4 \times 9}^{D(49)} | \quad \overbrace{9^2}^{D(9)}$$

$$= 16 \underset{+}{\underbrace{| 7 2 |}} 8 1$$

$$= 16 / 80 / 1$$
$$= 2401$$

**Example** 18: Find the square of 465

**Solution**: The groups of numbers for **465** are–

$$= \quad \overbrace{4^2}^{D(4)} \quad \overbrace{2 \times 4 \times 6}^{D(46)} \quad \overbrace{2 \times 4 \times 5 + 6^2}^{D(465)} \quad \overbrace{2 \times 6 \times 5}^{D(65)} \quad \overbrace{5^2}^{D(5)}$$

$$= 16 \underset{+}{\underbrace{| 4 8 |}} \underset{+}{\underbrace{7 6 |}} \underset{+}{\underbrace{6 0 |}} 2 5$$

$$= 20 / {}_1 5/ {}_1 2/2/5$$
$$= 216225$$

**Example** 19: Find the square of 687

**Solution**: The groups of numbers for **687** are–

$$D(6) / D(68) / D(687) / D(87) / D(7)$$
$$= 6^2 / 2 \times 6 \times 8 / 2 \times 6 \times 7 + 8^2 / 2 \times 8 \times 7 / 7^2$$
$$= 36 / {}_9 6 / {}_{14} 8 / {}_{11} 2 / {}_4 9$$
$$= 45/ {}_2 0/ {}_1 9/6/9$$
$$= 471969$$

**Example** 20: Find the square of 8254

**Solution**: The groups of numbers for 8254 are

**8, 82, 825, 8254, 254, 54 and 4.**
Duplex of 8 = $8^2$ = 64
Duplex of 82 = 2 x 8 x 2 = 32

Duplex of 825 = 2 x 8 x 5 + $2^2$ = 84
Duplex of 8254 = 2 x 8 x 4 + 2 x 2 x 5 = 84
Duplex of 254 = 2 x 2 x 4 + $3^2$ = 41
Duplex of 54 = 2 x 5 x 4 = 40
Duplex of 4 = $4^2$ = 16
Arrange the value of duplex as follows–

64 | $_3$2 | $_8$4 | $_8$4 | $_4$1 | $_4$0 | $_1$6
= 6 8 1 2 8 5 1 6

Hence $(8254)^2$ = 68128516

**Example 21:** Find the square of 4856

**Solution:** The groups for 4856 are:

**4, 48, 485, 4856, 856, 56, and 6**
Duplex of 4 = $4^2$ = 16
Duplex of 48 = 2 x 4 x 8 = 64
Duplex of 485 = 2 x 5 x 4 + $8^2$ =104
Duplex of 4856 = 2 x 4 x 6 + 2 x 8 x 5 =128
Duplex of 856 = 2 x 8 x 6 + $5^2$ =121
Duplex of 56 = 2 x 5 x 6 = 60
Duplex of 6 = $6^2$ = 36

Arrange the duplex of each number in digit separator.
= 16 / $_6$4 / $_{10}$4 / $_{12}$8 /$_{12}$1 / $_6$0 / $_3$6
= 23580736

**Example 22:** Find the square of 45612

**Solution:** The groups for 45612 are:

**4, 45, 456, 4561 , 45612, 5612, 612, 12 and 2**
Duplex of 4 = $4^2$ = 16
Duplex of 45 = 2 x 4 x 5 = 40
Duplex of 456 = 2 x 4 x 6 + $5^2$ =73
Duplex of 4561 = 2 x 4 x 1 + 2 x 5 x 6 = 68
Duplex of 45612 = 2 x 4 x 2 + 2 x 5 x 1 + $6^2$ =62

Duplex of 5612 = 2x 5 x 2 + 2 x 6 x 1 = 32
Duplex of 612 = 2 x 6 x 2 + $1^2$ = 25
Duplex of 12 = 2 x 1 x 2 = 4
Duplex of 2 = $2^2$ = 4

Arrange the duplex of each number in digit separator.
= 16 / $_4$0 / $_7$3 / $_6$8 / $_6$2 / $_3$2 / $_2$5 / 4 / 4
= 2080454544

Proceeding in the same fashion, readers can find the square of any number of their choice by using the Duplex Method. All the methods discussed above have their own beauty and it is now the readers who have to decide which method is the best according to a particular situation. The squaring of a number by the Urdhva-Tiryak sutra is the simple multiplication of a number twice; therefore it is left upto the readers to decide which method they want to adopt.

# 8

# Square Root

## Introduction

In mathematics, a **square root** of a number $x$ is a number $r$ such that $r^2 = x$, or, in other words, a number $r$ whose square (the result of multiplying the number by itself, or $r \times r$) is $x$. For example, 4 is a square root of 16 because $4^2 = 16$. Mathematically, if $x^2 = y$ then $x = y^{1/2} = \pm \sqrt{y}$

If you square 2, you get 4, and if you "take the square root of 4", you get 2; if you square 3, you get 9, and if you "take the square root of 9", you get 3.

$$2^2 = 4, \text{ so } \sqrt{4} = 2$$
$$3^2 = 9, \text{ so } \sqrt{9} = 3$$

Generally, extracting the square root of a number is considered a tedious job. We do have two sets of methods taught in our present day classroom:

a)  Factor Method

b)  Long Division Method.

Both the methods are lengthy and time consuming. The Vedic sutra helps us to find the square root of an exact square, merely by observation. Before we move to the Vedic Methodology to find the square root of a number, let us understand the following fundamental rule.

- A perfect square ends in 0, 1, 4, 5, 6 and 9
- A number is not a perfect square if it ends with 2, 3, 7 or 8

- If the given number has n digits then its square root will have n/2 digits if n is even else (n + 1)/ 2 digits if n is odd

Let us study the following square root table

**Table 1**

| N | $N^2$ | Last digit of $N^2$ | Digit sum of square number |
|---|---|---|---|
| 1 | 1 | 1 | 1 |
| 2 | 4 | 4 | 4 |
| 3 | 9 | 9 | 9 |
| 4 | 16 | 6 | 7 |
| 5 | 25 | 5 | 7 |
| 6 | 36 | 6 | 9 |
| 7 | 49 | 9 | 4 |
| 8 | 64 | 4 | 1 |
| 9 | 81 | 1 | 9 |
| 10 | 100 | 0 | 1 |

From the above table we conclude that –

1. A Complete square ending in 1 must have either 1 or 9 as the last digit of the square root.
2. A square ending in 4 must have 2 or 8 as the last digit of the square root.
3. A square ending in 6 must have 4 or 6 as the last digit of the square root.
4. A square ending in 5 will have 5 as the last digit of the square root.
5. A square ending in 9 must have 3 or 7 as the last digit of the square root.
6. A square ending in 00 will have 0 as the last digit of the square root.

7. Apart from the above, let us look at the following nearest square root table that will help us to find the square root of a number instantly.

### Table 2

| Number | Nearest Square Root | Number | Nearest Square Root |
|--------|---------------------|--------|---------------------|
| 1–3 | 1 | 4–8 | 2 |
| 9–15 | 3 | 16–24 | 4 |
| 25–35 | 5 | 36–48 | 6 |
| 49–63 | 7 | 64–80 | 8 |
| 81–99 | 9 | | |

## Vedic Method

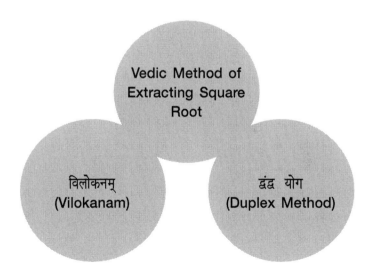

Vedic Method of Extracting Square Root

विलोकनम् (Vilokanam)

द्वंद्व योग (Duplex Method)

## Meaning of Vedic Sutra:

1. विलोकनम् (Vilokanam): It means, "by mere observation". This Vedic sutra will help you to find the square root

of 3–4 digits in 2–3 seconds, merely by observing the above two tables. The first table will help you to find the unit digit of the exact square root, whereas Table 2 will help you to find the ten's digit.

2. द्वंद्व-योग (Duplex method): A detailed description of this method can be found in this book itself in the chapter on Squares. This Vedic sutra is applicable to all, whether the given number is a perfect square root or not.

## Exact square root of 3–4 digits by Vilokanam (विलोकनम्) method

**Rule:**

- Make a group of two starting from the right
- Look at the unit digit of the number and observe your answer in Table 1. This will help you to decide the digit at the unit place
- Now move to the second group and find the ten's digit of your square root

**Example 1:** Find the square root of 2116.

**Solution:**

- Make a group of two from right.

$$\underline{\quad 21 \quad} \qquad \underline{\quad 16 \quad}$$
$$2^{nd} \text{ group} \qquad 1^{st} \text{ group}$$

- The unit digit of the $1^{st}$ pair is 6, so the square root ends in 4 or 6 (see Table 1)
- Now look at the second group. Since $16 < 21 < 25$, the digit at the ten's place = 4 (see Table 2)
- Now we have two options $\sqrt{2116} = 44$ or $46$

We know that $45^2 = 2025$, since $2116 > 2025$, therefore the desired square root will be more than 45.

Hence, $\sqrt{2116} = 46$

**Example 2:** Find the square root of 5184.

**Solution:**

- Make a group of two from the right.

$$\underline{\quad 51 \quad} \qquad \underline{\quad 84 \quad}$$

$2^{nd}$ **group**     $1^{st}$ **group**

- The unit digit of the $1^{st}$ pair is 4, so the square root ends in 2 or 8 (see Table 1)
- Now look at the second group. Since $49 < 51 < 84$, the digit at the ten's place = 7 (see Table 2)
- Now we have two options   $\sqrt{5184} = 72$ or 78

We know that $75^2 = 5625$, since $5184 < 5625$, therefore the desired square root will be less than 75

Hence, $\sqrt{5184} = 72$

**Example 3:** Find the square root of 9216.

**Solution:**

- Make a group of two from the right.

$$\underline{\quad 92 \quad} \qquad \underline{\quad 16 \quad}$$

$2^{nd}$ **group**     $1^{st}$ **group**

- The unit digit of $1^{st}$ pair is 6, so the square root ends in 4 or 6 (see Table 1)
- Now look at the second group. Since $81 < 92 < 100$, the digit at the ten's place = 9 (see Table 2).
- Now we have two options   $\sqrt{9216} = 94$ or 96

We know that $95^2 = 9025$, since $9216 > 9025$, therefore the desired square root will be more than 95.

Hence, $\sqrt{9216} = 96$.

**Example 4:** Find the square root of 676.

**Solution:**

- Make a group of two from the right.

$$\underline{\quad 06 \quad} \qquad \underline{\quad 76 \quad}$$

$$2^{nd} \text{ group} \qquad 1^{st} \text{ group}$$

- The unit digit of $1^{st}$ pair is 6, so the square root ends in 4 or 6 (see Table 1)
- Now look at the second group. Since $4 < 6 < 9$, the digit at the ten's place = 2 (see Table 2)
- Now we have two options $\sqrt{676} = 24$ or 26

We know that $25^2 = 625$, since $676 > 625$ therefore the desired square root will be more than 25

Hence, $\sqrt{676} = 26$.

## Square Root of 5–6 digits by the Vilokanam Method

**Rule:**
- Make a group of two digits starting from the right. Here we will have three groups. Denote the left digit by L, the middle digit by M and the right digit by R
- The first (L) and third (R) group will give us the hundred's place digit and the unit place digit. These two can be written only through observation, with the help of Table 1 and Table 2.
- Subtract $L^2$ from the $1^{st}$ pair and carry down the next digit from the dividend, as done in simple division.

- Compare the new dividend by 2 L M. Put different value of M in 2LM and select the best possible digit, so that 2 L M ≤ new dividend.
- In order to avoid confusion over the choice of the number, use the Casting out Nines rule.
- Exact square root = L M R

**Example 1:** Find the square root of 692224.

**Solution:**

Make a group of two digits, starting from the right.

$$\underline{69} \quad \underline{22} \quad \underline{24}$$

L        M        R

- Since the unit digit of the given number ends with 4, therefore the square root ends in 2 or 8 (Table 1).
- 69 in the left group lies between $8^2 < 64 < 9^2$, hence, L = 8 (Table 2).
- Now subtract $L^2$ from the given number and carry down the next digit from the dividend as shown here –

$$
\begin{array}{ccc}
69 & 22 & 24 \\
-\,8^2 & & \\
\hline
5 & 2 &
\end{array}
$$

→ carry down the next digit from the dividend
→ new dividend

- Compare the new dividend by 2 L M = 2 x 8 x M = 16 M. Put different values of M.
  For M = 3, 16 x 3 = 48 < 52
  and For M = 4 , 64 > 52. Hence, M = 3
- Now we are left with two options:
  $$\sqrt{69\ 22\ 24} = 8\ 3\ 2 \text{ or } 8\ 3\ 8$$
- Apply the Casting out Nines method to overcome the confusion in selecting the correct answer.

| Digit sum of 69 22 24 | Digit sum of $(832)^2$ | Digit sum of $(838)^2$ |
|---|---|---|
| 6 + 9 + 2 + 2 + 2 + 4 = 7 | 7 | 1 |

Hence, $\sqrt{69\ 22\ 24} = 832$

**Example 1:** Find the square root of 103041.

**Solution:**

- Make a group of two digits, starting from the right.

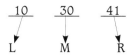

- Since the unit digit of the given number ends with 1, the square root ends in 1 or 9 (Table 1)
- 10 in the left group lies between $3^2 < 10 < 4^2$, hence, L = 3 (Table 2)
- Now subtract $L^2$ from the given number and carry down the next digit from the dividend as shown here –

10          30     41
$-\ 3^2$        |——————→ carry down the next digit from the dividend
1           3 ——————→ new dividend

- Compare the new dividend by 2 L M = 2 x 3 x M = 6 M. Put different values of M.
  For M = 2, 6 x 2 = 12 < 13
  and For M = 3 , 18 > 13. Hence, M = 2
- Now we are left with two options:–
  $\sqrt{10\ 30\ 41} = 321$ or 329
- Apply the Casting out Nines method to overcome the confusion in selecting the correct answer.

| Digit sum of 10 30 41 | Digit sum of $(321)^2$ | Digit sum of $(329)^2$ |
|:---:|:---:|:---:|
| 0 | 0 | 7 |

Hence, $\sqrt{10\ 30\ 41} = 321$

## द्वंद्व-योग विधि (Duplex Method)

**Rule:**

- The given number is first arranged in two-digit groups from right to left, and a single digit if any left over at the left hand end, is counted as a simple group by itself.
- The number of digits in the square root will be the same as the number of digit-groups in the given number itself including a single digit if any such there is.
- If a square root contains n digits, the square must consist of 2n or 2n–1 digits.
- And conversely, if the given number has n digits, the square root will contain n| 2 or (n+1)|2 digits.
- Group the number by placing a bar and put them in between the horizontal and vertical line bar, as shown in the given examples.

### Working procedure of the Duplex Method

The working of the Duplex method is as simple as straight division. Let us take a few examples to understand the modus operandi of this method.

**Example:** Find the square root of 529 by using the Duplex method.

**Solution:**

- Group the number $\overline{5}\ \overline{29}$ by placing a bar over it.
- Put a horizontal and vertical line as shown below.

|     | 5 | 29 |   |
|:---:|:---:|:---:|:---:|
| Q   |   |    |   |

- Since $2^2 < 5 < 3^2$, the first digit of the square root in the quotient column is 2. Double the quotient and set this down as divisor

| 4 | 5 | 29 |
|---|---|----|
| Q | 2 |    |

Remainder = $5 - 2^2$ = 1

- Put this remainder below the next dividend digit. Hence the next gross dividend = 12

| 4 | 5 | ₁29 |
|---|---|-----|
| Q | 2 | 3   |

- Divide 12 by 4 and put the quotient 3 in the quotient column.

$$12 \div 4 = 3$$

- Next dividend = 9
  Subtract the square of the quotient from the next dividend.
  Net dividend = $9 - 3^2$ = 0
  Since no more digits are left, the square root of 529 is 23.

**Example:** Find the square root of 4225

**Solution:**

Group the number $\overline{42}\ \overline{25}$ by placing a bar over it.
Put a horizontal and vertical line as shown below.

|   | 42 | 25 |
|---|----|----|
| Q |    |    |

- Since $6^2 < 12 < 7^2$, the first digit of the square root in the quotient column is 6. Double the quotient and set this down as divisor.

| 12 | 42 | 25 |
|----|----|----|
| Q  | 6  |    |

Remainder $= 42 - 6^2 = 6$

- Put this remainder below the next dividend digit. Hence the next gross dividend $= 62$.

| 12 | 42 | $_6$25 |
|----|----|--------|
| Q  | 6  |        |

- Divide 62 by 12 and put the quotient 5 in the quotient column and the remainder 2 below 5.

| 12 | 42 | $_6$2 | $_2$5 |
|----|----|-------|-------|
| Q  | 6  | 5     |       |

- Next dividend $= 25$    Subtract the square of the quotient from the next dividend.
  Net dividend $= 25 - 5^2 = 0$
  Since no more digits are left, the square root of 4225 is 65.

**Example**: Find the square root of 20736

**Solution**:

- Group the number 2 07 36, by placing a bar over it.
- Put horizontal and vertical line as shown below.

|   | 2 | 07 | 36 |
|---|---|----|----|
| Q |   |    |    |

- Since $1^2 < 2 < 2^2$, the first digit of the square root in the quotient column is 1. Double the quotient and set this down as divisor.

| 2 | 2 | 07 | 36 |
|---|---|----|----|
| Q | 1 | 4  |    |

Remainder $= 2 - 1^2 = 1$.

- Put this remainder below the next dividend digit. Hence the next gross dividend = 10

| 2 | 2 | $_1$07  36 |
|---|---|---|
| Q | 1 | |

- Divide 10 by 2
  $10 \div 2 = 5$, Q = 5 and R = 0

Since the remainder cannot be taken as 0 until the whole operation is completed, we need to take the quotient as less than 5. This is because we cannot consider the remainder zero in the middle of division.

Hence the **Revised Quotient = 4** and **Revised Remainder = 2**

| 2 | 2 | $_1$0 $_2$7  36 |
|---|---|---|
| Q | 1 | 4 |

- Next dividend = 27
  Subtract the square of quotient from the next dividend.
  Net dividend = $27 - 4^2 = 11$
  Divide 11 by 2 and write the quotient ( Q = 5) and remainder (R = 1) at their respective places.

| 2 | 2 | $_1$0 $_2$7  $_1$36 |
|---|---|---|
| Q | 1 | 4  5 |

- Gross dividend = 13
  Net dividend = 13 − (Duplex of 45)
  $$= 13 - 2 \times 4 \times 5$$
  $$= 13 - 40 = -27 < 0$$

(Since Net dividend is less than zero, we can't take the quotient (Q= 5) as taken above. Now for $11 \div 2$, **Revised Q = 4** and **Revised R = 3**.)

| 2 | 2 | $_1$0 $_2$7  $_3$36 |
|---|---|---|
| Q | 1 | 4  4 |

- Gross dividend = 33

  Net dividend = 33 – (Duplex of 44)

  = 33 – 2 x 4 x 4

  = 33 – 32

  = 1

Divide 1 by 2 and put the quotient **Q** = 0 and remainder **R** = 1 in its proper place.

| 2 | 2 | $_10$ $_27$ $_13$ $_16$ |
|---|---|---|
| Q | 1 | 4   4.0 |

- Next gross dividend = 16
- Net dividend = 16 – duplex of 4

  $= 16 - 4^2$

  $= 0$

Since no more digits are left, the square root of 20736 is 144

**Example**: Find the square root of 25747576

**Solution**:

| 10 | 25 | $_74$ $_45$ $_54$ $_57$ $_16$ |
|---|---|---|
| Q | 5 | 0  7  4.000 |

A step by step illustration has been done here for your convenient.

- Number of digits in the square root = 4
- The first digit of the square root = 5
- Remainder = $25 - 5^2$ = 0, place it below 7.
- Next Dividend = 07 and Divisor = 2 x 5 = 10
- For 07 ÷10 , Q = 0 and R = 7, place it at proper place.
- Next dividend = 74 and corrected dividend

  $= 74 - 0^2 = 74$
- For 74 ÷10 , Q = 7 and R = 4
- Next dividend = 45 and corrected dividend

  = 45 – 2 x 0 x7 = 45
- For 45÷10 , Q= 4 and R = 5

- Since we have so far got 4 digits in the quotient column, the perfect square root is obtained. The next digit in the quotient column will give the remainder, if any.
- Remainder= $54 - 2 \times 0 \times 4 - 7^2 = 5$
- Next dividend = 57 and remainder
  $= 57 - 2 \times 4 \times 7 = 1$
- Next dividend =16 and the last remainder
  $= 16 - 4^2 = 0$
- I think you have understood the Duplex method, therefore the next two examples given below are without much detail, though a brief description is provided here for your convenience.

**Example**: Find the square root of 45 31 98 24

**Solution**:

| 12 | 45 | $_9 3 \ _9 1 \ _6 9 \ _3 8 \ _1 2 \ 4$ |
|----|----|------------------------------------------|
| Q  | 6  | 7 3 2 .000 |

A step by step illustration is done here for your convenient.

- Number of digits in the square root = 4
- The first digit of square root = 6
- Remainder = $45 - 6^2 = 9$, place it below 3.
- Next Dividend = 93 and Divisor = 2 x 6 = 12
- For 93 ÷12, Q = 7 and R = 9, place it at the proper place.
- Next dividend = 91 and corrected dividend
  $= 91 - 7^2 = 42$
- For 42 ÷12 , Q = 3 and R = 6
- Next dividend = 69 and corrected dividend
  $= 69 - 2 \times 7 \times 3 = 27$
- For 27÷12 , Q= 2 and R = 3
- Since we have so far got 4 digits in the quotient column, the perfect square root is obtained. The next digit in the quotient column will give the remainder if any.

- Remainder= $38 - 2 \times 7 \times 2 - 3^2 = 1$
- Next dividend = 12 and remainder
  $= 12 - 2 \times 3 \times 2 = 0$
- Next dividend = 4 and the last remainder
  $= 4 - 2^2 = 0$

**Example**: Find the square root of 52443907 up to 1 decimal place.

**Solution**:

| 14 | 52 | $_34$ $_64$ $_43$ $_{13}9$ $_70$ 7 |
|----|----|---------------------------|
| Q  | 7  | 2 4 1. 8                  |

A step by step illustration is done here for your convenient.

- Number of digits in the square root = 4
- The first digit of square root = 7
- Remainder = $52 - 7^2 = 3$, place it below 4.
- Next Dividend = 34 and Divisor = $2 \times 7 = 14$
- For $34 \div 14$, Q = 2 and R = 6, place it at proper place.
- Next dividend = 64 and corrected dividend
  $= 64 - 2^2 = 60$
- For $60 \div 14$, Q = 4 and R = 4
- Next dividend = 43 and corrected dividend
  $= 43 - 2 \times 2 \times 4 = 27$
- For $27 \div 14$, Q= 1 and R = 13
- Since we have so far got 4 digits in the quotient column, the perfect square root is obtained. The next digit in the quotient column will give the digit after decimal.
- Remainder= $139 - 2 \times 2 \times 1 - 4^2 = 119$
- Next dividend = 119 and Quotient = $119 \div 14 = 8$ and Remainder = $119 - 112 = 7$

# 9

# Square Root of Irrational Number

## Introduction

In the previous chapter we learnt the square root of a perfect number, but suppose you are in a situation where you need to find the square root of an irrational number? Now the big question is – how will you extract the square root of such a number? You may find yourself in such a situation while solving the problem of surds in arithmetic, mensuration or trigonometry. The conventional method of finding the square root of such a number is possible only through Long Division method and this method is time consuming and error prone in this case. Here is a method that helps you to find the square root of the irrational by mere observation, and with a little practice; you will be able to find the square root of such numbers mentally.

## Vedic Method

There is no such reference of this method in the Vedic mathematics book written by Bharti Krishna Tirthaji Maharaj, but I have used the theory of Differential Calculus and the Vedic sutra, Vilokanam (विलोकनम्) in extracting the square root of such a number. The best part of this method is that we are free to choose the number of digits after the decimal. Generally, in competitive examinations, we use the technique of approximation in finding the square root of such numbers and take a maximum of two digits after the decimal place.

**Square root of Irrational Number =**

$$\sqrt{\text{Nearest Perfect square}} + \frac{\text{Deviation from Irrational number}}{2 \times \sqrt{\text{Nearest Perfect square}}}$$

Let us take few examples to understand the modus operandi of extracting the square root of a number that is not a perfect square.

**Example 1: Find the square root of 79**

**Solution**: Perfect square approaching 79 is 81.

$$\text{Deviation} = 79 - 81 = -2$$

$$\sqrt{79} = \sqrt{81} - \frac{2}{2 \times \sqrt{81}}$$

$$= 9 - \frac{1}{9} = 9 - 0.999 = 8.001$$

**Example 2**: Find the square root of 174

**Solution**: Perfect square approaching 174 is 169

$$\text{Deviation} = 174 - 169 = 5$$

$$\sqrt{174} = \sqrt{169} + \frac{5}{2 \times \sqrt{169}}$$

$$= 13 + \frac{5}{26}$$

$$= 13.192$$

**Example 3**: Find the square root of 474

**Solution**: Perfect square approaching 474 is 484.

$$\text{Deviation} = 474 - 484 = -10$$

$$\sqrt{474} = \sqrt{484} - \frac{10}{2 \times \sqrt{484}}$$

$$= 22 - \frac{5}{22}$$

$$= 22 - 0.227 = 21.773$$

**Example 4:** Find the square root of 187

**Solution:** Perfect square approaching 187 is 196.

$$\text{Deviation} = 187 - 196 = -9$$
$$\sqrt{187} = \sqrt{196} \dfrac{-9}{2 \text{ x} \sqrt{196}}$$
$$= 14 - \dfrac{9}{28}$$
$$= 14 - 0.32$$
$$= 13.67$$

**Example 5:** Find the square root of 28

**Solution:** Perfect square approaching 28 is 25.

$$\text{Deviation} = 28 - 25 = 3$$
$$\sqrt{28} = \sqrt{25} \dfrac{+3}{2 \text{ x} \sqrt{25}}$$
$$= 5 + \dfrac{3}{10}$$
$$= 5.3$$

**Example 6:** Find the square root of 34

**Solution:** Perfect square approaching 34 is 36.

$$\text{Deviation} = 34 - 36 = -2$$
$$\sqrt{34} = \sqrt{36} \dfrac{-2}{2 \text{ x} \sqrt{36}}$$
$$= 6 - \dfrac{1}{6}$$
$$= 6 - 0.166 = 5.844$$

**Example 7:** Find the square root of 204

**Solution:** Perfect square approaching 204 is 196.

$$\text{Deviation} = 204 - 196 = 8$$
$$\sqrt{204} = \sqrt{196} \dfrac{+8}{2 \text{ x} \sqrt{196}}$$
$$= 14 + \dfrac{8}{28}$$
$$= 14 + 0.285$$
$$= 14.285$$

# Cube

## Introduction

When a number is multiplied by itself three times, the number so obtained is called the cube of that number. In general, a x a x a = $a^3$. Here is the cube of the first ten numbers.

| Number | 1 | 2 | 3 | 4 | 5 | 6 | 7 | 8 | 9 | 10 |
|--------|---|---|---|---|---|---|---|---|---|-----|
| Cube | 1 | 8 | 27 | 64 | 125 | 216 | 343 | 512 | 729 | 1000 |

Cubes of large numbers are not useful as far as the syllabus of CBSE, ICSE are concerned, though many questions based on cubing a number are asked in competitive examinations. In arithmetic too, while solving the problem of mensuration, we can't escape the tiring process of cubing the numbers. Cubing a large number is a tough task, as the traditional method only allows us to multiply the particular number three times or use the formula for binomial expansion. There is no room to experiment in the traditional method.

Traditional method of cubing:

$(988)^3$ = 988 x 988 x 988

```
         9 8 8
       X 9 8 8
       7 9 0 4
     7 9 0 4 x
   8 8 9 2 x x
   9 7 6 1 4 4
       X 9 8 8
   7 8 0 9 1 5 2
 7 8 0 9 1 5 2 x
 8 7 8 5 2 9 6 x x
 9 6 4 4 3 0 2 7 2
```

There is another traditional method which can be termed as better than the above method. The binomial expansion of $(a + b)^3$ and $(a - b)^3$ will reduce calculation time a bit but this method is still not suitable as far as its application in competitive examinations is concerned.

- $(988)^3 = (1000 - 12)^3 = (1000)^3 - 3 (1000)^2 \times 12 + 3 \times 1000 \times (12)^2 - (12)^3$

  $= 1000000000 - 36000000 + 432000 - 1728$

  $= 964430272$

  [Note: Applying the binomial expansion

  $(a-b)^3 = a^3 - 3a^2b + 3ab^2 - b^3$ ]

- $(108)^3 = (100 + 8)^3 = (100)^3 + 3 \times (100)^2 \times 8 + 3 \times 100 \times (8)^2 + (8)^3$

  $= 1000000 + 240000 + 19200 + 512$

  $= 1259712$

  [Applying binomial expansion

  $(a-b)^3 = a^3 + 3a^2b + 3ab^2 + b^3$]

Vedic mathematics presents several very interesting methods to find the cube of any number in a few seconds. Unlike the traditional method, it is easy, interesting and short. It reduces the time by $1/10^{th}$ and that's why the Vedic method is far superior to the traditional method of finding cubes.

## Vedic Sutra for cubing a number

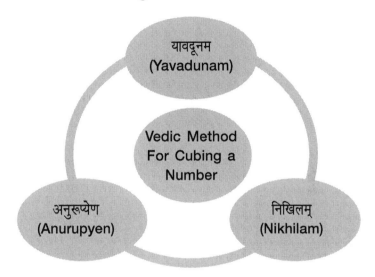

### 1. Yavadunam (याबदूनम)

This formula works better when the number to be cubed is near the base. The base should be in the form of $10^n$, where n is a natural number. This formula has limited application.

### Working Rule

- Check whether the number is near the base 10, 100, 1000... or not
- Find the excess or deficit number from the base.
- The whole operation is to be performed in three parts. In the $1^{st}$ part, add twice the excess/ deficit to the original number.
- $2^{nd}$ part = New Excess (Number obtained in $1^{st}$ part – base) x original excess/deficit
- $3^{rd}$ part = cube of excess.

Mathematically, if a = original number and d = deviation (Excess/ Deficit from the Base) then the whole operation can be summed up as —

$a^3$ = a+2d / [ (a + 2d ) – base ] x d / $d^3$

let us take some example.

**Example**: Find the cube of 12

**Solution**: Here, in $(12)^3$ ; a = 12 (original number) is nearer to the base 10. Hence, base = 10 and Excess = + 2.

$1^{st}$ part = a + 2d = 12 + 2 x 2 = 16
$2^{nd}$ part = (16 – 10) x 2 = 12
$3^{rd}$ part = $2^3$
Combining all the three parts we get,
$(12)^3$ = 16 | 12 | 8
       = 17 | 2 | 8
       = 1728

**Example**: Find the cube of 96

**Solution**: Here, in $(96)^3$ ; a = 96 (original number) is nearer to the base 100. Hence, base = 100 and deficit = – 4

$1^{st}$ part = a + 2d = 96 – 2 x 4 = 88
$2^{nd}$ part = (88 – 100) x –4 = 48
$3^{rd}$ part = $(–4)^3$ = – 64
Combining all three parts we get,
$(96)^3$ = 8 8 | 4 8 | −64
       = 88 |47 + 1 | −64
       = 88 | 47 | 100 – 64
       = 88 | 47 | 36

($3^{rd}$ part is negative, so add 100 to the negative part to make it 100 – 64 = 36. Subtract 1 from the previous part, thus making 48 to 47.)

**Example**: Find the cube of 105

**Solution**: a = 105, Base = 100. Excess = 5

$1^{st}$ part = a + 2d = 105 + 2 x 5 = 115
$2^{nd}$ part = (115 – 105) x 5 = 50
$3^{rd}$ part = $5^3$ = 125

Combining all three parts, we get,

$(105)^3$ = 115 | 7 5 | 1 25
            = 115| 75 + 1 | 25
            = 115 | 76 | 25

(The number of digits in each part depends on the number of zeros in the base digit. If the base is taken as 100, the number of digits permissible in each part is 2.)

## 2. Anurupyen (अनुरूप्येण):

Anurupyen Vedic sutra is based on the concept of Geometric Progression. In a geometric series, every next number is the multiple of some constant ratio called **r**, or the common ratio.
If a = first term, r = common ratio, then the $n^{th}$ term $(t_n)$ = a $r^{n-1}$.
If a, b, c are in Geometric series then,
Second term/ first term = third term/ second term = r

b / a = c / b
or, $b^2$ = ac

Here are a few examples of geometric series.

   i) 2, 8, 32 ...
   ii) 5, 25, 125, 625,...

In the first example (i), a = first term and r = common ratio = 8/2 = 4

   $T_n$ = 2 ( 4 )$^{n-1}$ = 2. $2^{2(n-1)}$

In the second example, a = 5 and r = 5, hence $T_n$ = 5 $(5)^{n-1}$

**Rule**:

* First, take the cube of the first digit (a) and multiply it with the common ratio (b/a), in a row of 4 figures

- Double the second and the third number and put it down, under the second and third numbers. Finally add up the two rows

This can be further clarified with the help of this table

|  | $1^{st}$ | $2^{nd}$ | $3^{rd}$ | $4^{th}$ |
|---|---|---|---|---|
| $(ab)^3 =$ | $a^3$ | $a^2b$ | $ab^2$ | $b^3$ |
|  | $+$ | $2a^2b$ | $2ab^2$ |  |
|  | $a^3$ | $3a^2\ b$ | $3\ a\ b^2$ | $b^3$ |

In the binomial expansion we know,
$(a + b)^3 = a^3 + 3a^2b + 3ab^2 + b^3$ .

The Anurupyen method is just an extension of the above expansion. This can be simplified again if we take the help of geometric progression.

$$(ab)^3 = \quad a \quad ar \quad ar^2 \quad ar^3$$
$$+ \quad 2ar \quad 2ar^2 \qquad \text{(Where } r = b/a)$$
$$\overline{a \quad 3\ ar \quad 3ar^2 \quad a\ r^3}$$

Let us take a few examples to understand the basic modus operandi.

**Example:** Find the cube of 12?

**Solution:** $(12)^3 = ?$

Here, say **a** = 1 and **b** = 2
**r** = common ratio = b/a = 2/ 1 = 2
Hence the table arrangement will be as follows:–
$(12)^3 = $ 1      2      4      8
       +      4      8
       1      7      2      8

**Important points:**

1. *If you start with the cube of the first digit and multiply with the geometric ratio up to the next three numbers,*

*the 4th number of the series will be the cube of the 2nd digit.*

2. *Addition should be done from right to left, keeping only a single digit at a time and the remaining digit will be carried-over to the next column, and so on.*

**Example:** Find the cube of 15

**Solution:**    $( 15 )^3 = ?$

Here say **a = 1** and **b = 5**

r = common ratio = b/a = 5 / 1 = 5

Hence the table arrangement will be as follows

$(15)^3 = $  1      5      25      125

\+          10      50

1      *1* 5    *7* 5    *1 2* 5

= 3 3 75

The excess digit, leaving the unit digit from each column (from right to left) is transferred to the next column. In the above example, the excess digit is underlined.

**Example:** Find the cube of 19

**Solution:**    $( 19 )^3 = ?$

Here say **a = 1** and **b = 9**

r = common ratio = b/a = 9/ 1 = 9

Hence the table arrangement will be as follows:-

1      9      81      729

        18      162

1      *2* 7    *24* 3    *72* 9

= 6 8 5 9

**Example**: Find the cube of 32

**Solution**: $(32)^3 = ?$

Here, say $a = 3$ and $b = 2$

r = common ratio = b/a = 2/3

Hence the first line of the table arrangement will be as follows:-

$(32)^3 = 3^3$ (27)   27 x 2 / 3 (18)   18 x 2 / 3 (12)   12 x 2 / 3 (8)

| | | | | |
|---|---|---|---|---|
| = | 27 | 18 | 12 | 8 |
| | | 36 | 24 | |
| = | 2 7 | 5 4 | 3 6 | 8 |

=   3 2 7 6 8

**Example**: Find the cube of 46

**Solution**: $( 46 )^3 = ?$

Here say $a = 4$ and $b = 6$

r = common ratio = b/a = 6 / 4

Hence ,the table arrangement will be as follows:-

$(46)^3 = 4^3$ (=64)   64 x 6/4 (=96)   96 x 6/4 (=144)
144 x 8/9 (=512)

| | | | | |
|---|---|---|---|---|
| | 64 | 96 | 144 | 216 |
| | | 192 | 288 | |
| | 64 | 288 | 432 | 216 |
| = | 64 | 288 | 432 | 216 |
| = | 64 | 288 | 453 | 6 |
| = | 64 | 333 | 3 | 6 |

= 97336

**Example**: Find the cube of 105

**Solution**: $(105)^3 = ?$

Here a = 10 and b = 5, hence common ratio = 5 / 10 = 1/ 2

$$(105)^3 = \begin{array}{rrrr} 1000 & 500 & 250 & 125 \\ + & 1000 & 500 & \\ \hline 1000 & 1500 & 750 & 125 \end{array}$$

Simple addition as done in the above example will give you wrong answer. Now the big question is what next?-

The answer is very simple. If the number is between 100 – 999, put 1, 2, and 3 zeros after each digit as shown here-

$$1000000 + 150000 + 7500 + 125 = 1157625$$

In case the number is above 1000, you need to put 2, 4 and 6 zeros after each digit. Let us find the cube of a number above 1000.

**Example**: Find the cube of 1001

**Solution**: $(1001)^3 = ?$

Here, a = 10 and b =01, hence common ratio = 01 / 10 = 1/ 10

$$(1001)^3 = \begin{array}{rrrr} 1000 & 100 & 10 & 1 \\ + & 200 & 20 & \\ \hline 1000 & 300 & 30 & 1 \end{array}$$

Since the number is above 1000, start putting 6, 4 and 2 zeros from the extreme left before adding. Once the process of putting zeros get complete, you can simply add to get the result.

$(1001)^3 = 1000000000 + 3000000 + 3000 + 1 = 1003003001$

## 3. Nikhilam (निखिलम्) Vedic Sutra

### Rule

The modus operandi of the Nikhilam Sutra can be described in the following steps:-

- First, take the deviation of the number to be cubed from its base. The base should be the multiple of 10. If the base is 10 , 100,1000------ then sub-base = 1; on the other hand, if the base = 40, then the sub- base = 4, because 40 = 4 x 10,

- The whole cubing process then involves 3 steps.
  A) (Number to be cubed + 2 x deviation from the base) x (sub-base)$^2$
  B) {3 x (deviation)$^2$} x sub-base
  C) (Deviation)$^3$
- If there is no sub-base, then the calculation becomes very easy.

Let us take some example to understand the modus-operandi of this method.

**Example**: Find the cube of 25 using Nikhilam Sutra

**Solution**: 25 is nearer to the base 20 ( 2 x 10), hence–

$$\text{Deviation} = 25 - 10 = 5, \text{ sub- base} = 2$$
$$= (25 + 2 \times 5) \times 2^2 \mid (3 \times 5^2) \times 2 \mid 5^3$$
$$= 140 \mid 150 \mid 125$$
$$= 140 \mid 162 \mid 5$$
$$= 15625$$

**Example**: Find the cube of 58 by using Nikhilam Sutra

**Solution**: 58 = 5 x 10 + 8

base = 10      sub- base = 10      and excess = 8

Hence $(58)^3 = [\ \underline{58 + 2 \times (8)}\ ] \times 5^2 \mid \underline{3 \times (\ 8)^2 \times 5}\ \mid (8)^3$

$\qquad\qquad\qquad$ 1$^{st}$ term $\qquad$ 2$^{nd}$ term $\qquad$ 3$^{rd}$ term

$$= 1850 \mid 960 \mid 512$$
$$= 1850 \mid 1011 \mid 2$$
$$= 1951 \mid 1 \mid 2$$

Hence $(58)^3 = 195112$

**Example**: Find the cube of 98 by using Nikhilam Sutra?

**Solution**: 98 is nearer to the base 100

Deviation = 98–100 = –2

Hence $(98)^3$ = $\underbrace{98 + 2 \text{ x } (-2)}$ | $\underbrace{3 \text{ x } (-2)^2}$ | $\underbrace{(-2)^3}$

$1^{st}$ term      $2^{nd}$ term    $3^{rd}$ term

= 94 | 12 | –8

= 94 | 11 | 100–8

= 94 | 11 | 92

Hence $(98)^3$ = 941192

**Example**: Find the cube of 104 by using Nikhilam Sutra

**Solution**: Working base = 100

Deviation = 104 – 100 = 4

$(104)^3$ = 104 + 4 x 2 | 3 x $4^2$ | $4^3$

= 112 | 48 | 64

= 1124864

(Since the base =100, there should be 2 digits in each digit separator)

**Example**: Find the cube of 997 by using Nikhilam Sutra

**Solution**: Working base = 1000

Deviation = 997 – 1000 = – 3

$(997)^3$ = $\underbrace{997 + (-3) \text{ x } 2}$ | $\underbrace{3 \text{ x } (-3)^2}$ | $\underbrace{(-3)^3}$

$1^{st}$ part      $2^{nd}$ part    $3^{rd}$ part

= 991 | 27 | –27

= 991 | 026 | 1000 – 27

= 991 026 973

Since the base is 1000, there should be 3 digits in each digit separator. Therefore, 26 in the second part should be changed to 026. In the $3^{rd}$ part, there is a negative sign, so subtract it from the base 1000. Hence in the $3^{rd}$ part, we will have, 1000 – 27 = 973.This change will be adjusted by reducing 1 from the $2^{nd}$ part. Thus, 27 in the second part now become 27 – 1 = 26.

$(997)^3$ = 991026973

# 11

# Cube Root

## Introduction

The cube of a number is that number raised to the power 3. Thus, the cube of **a** is $a^3$ .

Suppose $a^3 = x$

$$\Rightarrow a = x^{1/3}$$

The present curriculum taught in our school uses only Factorization of a number and thus is a cumbersome technique. Moreover, it is a time consuming process and thus not fit for the competitive examinations. Suppose you are asked to find the cube root of a number having 10 digits, the present method taught in our school curriculum will take you 5–10 minutes, but if you are familiar with the Vedic method, it will take only 20–25 seconds to extract the cube root of even 8–9 digits. There are around 4 or 5 Vedic methods written in the original work of Sri Swami Krisna Tirthaji Maharaj.

Let us see one example.

**Example:** Find the cube root of 830584

**Solution:**

| | |
|---|---|
| 2 | 592704 |
| 2 | 296352 |
| 2 | 148176 |
| 2 | 74088 |
| 2 | 37044 |
| 2 | 18522 |
| 3 | 9261 |
| 3 | 3087 |
| 3 | 1029 |
| 7 | 343 |
| 7 | 49 |
| 7 | 7 |
| | 1 |

Hence, 592704 = 2 x 2 x 2 x 2 x 2 x 2 x 3 x 3 x 3 x 7 x 7 x 7

= 2 x 2 x 3 x 7

= 84

The Vedic method will instantly give you the answer 84 by mere observation.

The great Indian Astronomer and Mathematician Aryabhatta has also mentioned in his book, *Ganita Pada,* a method to extract the cube root of any number, but the method is too complex to understand. The fifth sloka of **Aryabhatta's** book **Ganita-Pada** reads as follows.

अघनाद भजेद द्वितीयात त्रिगुणेन घनस्य मूलवर्गेण ।
वर्ग स्त्रिपूर्व गुणितः शोध्यः प्रथमाद धनस्य घनात् ।।

Before Aryabhatta's, there was no proper method to extract the cube root of any number. All the mathematicians after him followed his method with some modifications here or there. This elegant method described by Aryabhatta has more symmetry with the Vedic Method used by Swami Bharathi Krishna Tirtathji Maharaj

Before we move further, look at the following table carefully. This table will help you to determine the unit digit of a cube root.

**Table 1**

| If a cube ends in | The unit digit of a cube root will be |
|---|---|
| 0 | 0 |
| 1 | 1 |
| 2 | 8 |
| 3 | 7 |
| 4 | 4 |
| 5 | 5 |
| 6 | 6 |
| 7 | 3 |
| 8 | 2 |
| 9 | 9 |

From the above table, we can conclude the following facts:-

*1, 4, 5, 6, 9, and 0 repeat themselves in the cube ending.*

*2, 3, 7 and 8 have an inter–play of complements from 10.*

The Left digit of a cube root having more than 7 digits, or the ten's digit of a cube root having less than 7 digits, can be extracted with the help of the following table.

**Table 2**

| Left-most pair of the cube root | Nearest cube root |
|---|---|
| 1 – 7 | 1 |
| 8– 26 | 2 |
| 27 – 63 | 3 |

| | |
|---|---|
| 64 – 124 | 4 |
| 125 – 215 | 5 |
| 216 – 342 | 6 |
| 343 – 511 | 7 |
| 512 – 728 | 8 |
| 729 – 999 | 9 |

There is another very important table that will help you to find the cube in ambiguous cases. The details will follow the next table.

**Table 3**

| Number | Cube | Beejank of cube |
|---|---|---|
| 1 | 1 | 1 |
| 2 | 8 | 8 |
| 3 | 27 | 0 |
| 4 | 64 | 1 |
| 5 | 125 | 8 |
| 6 | 216 | 0 |
| 7 | 343 | 1 |
| 8 | 512 | 8 |
| 9 | 729 | 0 |

The beejank (बीजांक) of a number is determined through the Casting out Nines (नवशेष) method described in the book. The Vedic method described here is effective as long as the number whose cube root is to be extracted is a perfect cube.

Now the bigger question is – *How will you determine that the given number is a perfect cube or not?*

The Vedic method has the answer of your question. Find the digit sum of that number by using the Casting out Nines method. If the digit sum of that number is found to be 0, 1 or 8 then the given number is a perfect cube.

**Example**: Is 1729 a perfect cube?

**Solution**: The digit sum of 1729 is $1 + 7 + 2 + 9 = 1$

Hence, it is a perfect cube

**Example**: Is 9528127 a perfect cube?

**Solution**: The digit sum of 9528127 is
$9 + 5 + 2 + 8 + 1 + 2 + 7 = 7$

Hence, it is not a perfect cube.

## Vedic Method

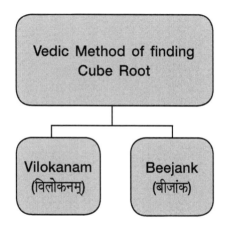

## Meaning of Vedic Sutra:

1. **Vilokanam** (विलोकनम्): The literal meaning of this sutra is – by inspection or observation. This sutra helps you to determine the unit digit and ten's digit of a perfect cube root in a few seconds, if the number whose cube root is to be extracted is less than 7 digits. Moreover, if the number whose cube root is to be extracted is more than 7 digits, this sutra will help you to find the L(Left digit) and R(Right digit) of the cube root.

2. **Beejank method** (बीजांक विधि): Many a time, while extracting the cube root of a number having more than 7 digits, you will get two results for M (Middle term) and will be little confused in deciding which one is the suitable value for M and the Beejank method will be there to help you out. The beejank of a number is extracted using the Casting out Nines method.

Let us now start with the Vedic method of finding cube root. For the convenience of readers, this chapter has been divided into two segments.

Case 1: Cube Root of a number having less than 7 digits
Case 2: Cube Root of a number having more than 7 digits, but less than 10 digits
Case 3: Cube Root of a number greater than 7 digits, but ending with even numbers.

**General rule to follow in all the three cases:**

1. Form the groups of three digits from right to left. The number of groups thus formed will give you an idea of the number of digits in its cube root.
   i.   A number having 4, 5 or 6 digits will have a cube root of 2 digits number.
   ii.  A number having 7, 8 or 9 digits will have a cube root of 3 digits number.
   iii. A number having 10, 11 or 12 digits will have a cube root of 4 digit numbers.

   **Example:** a) $\overline{3}$ $\overline{4\ 3\ 1}$ $\overline{3\ 3\ 1}$ has three groups, hence its cube root will contain 3 digits.

   b) $\overline{3\ 9}$ $\overline{3\ 0\ 4}$ has two groups, hence its cube root will have 2 digits.

2. Find the digit of the unit place by observing the ght-most pair and the left digit of the cube root by observing the left-most pair. Table 1 will help you to determine the

unit digit of cube root and Table 2 will give you the Left digit of the cube root.

*Case 1: Cube Root of a number having less than 7 digits.*

The Vedic method of extracting the cube root of a number having less than 7 digits is done by the Vilokanam ( विलोकनम्) method. Form the group of three digits from right to left as discussed above and find the unit digit and ten's digit with the help of Table 1 and Table 2.

**Example 1:** Find the cube root of 3375

**Solution:** The Vedic method as described above will have the following steps.

1. Place the bar over the number making the group
   $\overline{3}\ \overline{375}$
2. Since the bar is placed on two numbers, the cube root of this number contains two digits.
3. The first bar falls on unit digit 5, and with the help of Table 1 we can say that the unit digit of the cube root is 5.
4. The next bar falls on 3, and looking upon the Table 2 we see that the left digit of the cube root will be 1.

   Hence $\sqrt[3]{3375} = 15$.

**Example 2:** Find the cube root of 97336

**Solution:**

1. Place the bar over the number from right to left.
   $\overline{9\ 7}\ \overline{3\ 3\ 6}$
2. The cube ends with 6 hence the unit digit of cube root is 6 (see Table 1 for the reference)
3. The ten's digit of the cube root is 4, because
   $4^3 < 97 < 5^3$

   Hence, $\sqrt[3]{9\ 7\ 3\ 3\ 6} = 4\ 6$

**Example 3:** Find the cube root of 941192

**Solution:**

1.  Place the bar over the number from left to right, leaving two digits at a time.
    $$\overline{9\ 41}\ \overline{1\ 9\ 2}$$

2.  Since the bar is placed on two numbers, the cube root will contain only two digits.

3.  Since the unit digit of this number is 2, the unit digit of cube root is 8.

4.  For the ten's digit, take the least of
    $$9^3 < 941 < 10^3$$
    Ten's digit = 9

    Hence, $\sqrt[3]{941192} = 98$

*Case 2: Cube root of a number having more than 7 digits but less than 10 digits.*

The cube root of a number having more than 7 digits through the Vedic method needs a little practice and patience initially. A little practice and understanding of the fundamental concepts will help you to extract the cube root of much bigger number in less than 10 seconds. As discussed above, the cube root of more than 7 digit numbers will involve three digits, the unit digit (R) and the digit at the hundred's place (L) can be obtained quite easily with the help of Vilokanam (विलोकनम्) method, whereas the middle digit (M) will be determined with the help of the Beejank (बीजांक या नवशेष) method.

**Rule:**

1.  Denote the left digit by L, the middle digit by M and the right digit by R.
2.  Subtract $R^3$ from the number and cancel the last zero.
3.  The Middle digit of the cube root is obtained by 3 $R^2M$. Place a different value of M so that we may reach to the

unit digit of the number obtained in the previous step. In case you obtain more than one value of M, check by theBeejank method (Refer Table 3) which value of M is best suited in this case.

4.  The cube root of such a number = L M R.

**Example 4:** Find the cube root of 76765625

**Solution:**

1.  Group the number from right to left by placing the bar.

    $$\sqrt[3]{\overline{7\ 6}\ \overline{7\ 6\ 5}\ \overline{6\ 2\ 5}}$$

2.  The unit digit of the cube root is 5 (Refer Table 1). Hence R = 5.

3.  Moreover $4^3 < 76 < 5^3$, so L = 4 (Refer Table 2)

4.  Subtract $R^3$ from the number and forget the last digit (0) obtained after subtraction.

    $$\begin{array}{r} 7\ 6\ 7\ 6\ 5\ 6\ 2\ 5 \\ -1\ 2\ 5 \\ \hline 7\ 6\ 7\ 6\ 5\ 5\ 0\ \cancel{0} \end{array}$$

5.  Middle digit of the cube root is obtained by $3\ R^2\ M$.
    $3\ R^2\ M. = 3\ x\ 5^2\ x\ M = 75\ M$

6.  Now, the bigger question is what value of M should be put in 75 M, so that the unit digit of the result obtained is equal to the unit digit of 767550 obtained in step 3. We find here 4 options – they are 2, 4, 6 and 8. Hence, M = 2, 4, 6 or 8. M = 8 can be ruled out as 75 x 8 = 600 > 550. Still, we are left with 3 options.

7.  This situation can be handled by applying the Beejank method.
    Beejank of 76765625 = 8
    Beejank of $(425\ )^3$ = 8
    Beejank of $(445\ )^3$ = 1
    Beejank of $(465\ )^3$ = 0

This clearly shows that our answer is 425.

**Example 5:** Find the cube root of 84604519

**Solution:**

1.  Group the number by placing the bar as shown below. Since there are three groups, the cube root will have three digits.

    $$\sqrt[3]{8\ 4\ 6\ 0\ 4\ 5\ 1\ 9}$$

2.  The unit digit of the cube root is 9. Hence R = 9 (Refer Table 1)
3.  Moreover $4^3 < 84 < 5^3$, so L = 4 (Refer Table 2)
4.  Subtract $R^3$ from the number and forget the last digit (0) obtained after subtraction.

    $$\begin{array}{r} 8\ 4\ 6\ 0\ 4\ 5\ 1\ 9 \\ -7\ 2\ 9 \\ \hline 8\ 4\ 6\ 0\ 3\ 7\ 9\ 0\!\!\!/ \end{array}$$

5.  Middle digit of the cube root is obtained by 3 $R^2$ M.
    $R^2$ M. = 3 x $9^2$ x M = 243M
6.  The unit digit of the number obtained in step 5 is 9 so put M = 3 to obtain the Middle digit.
7.  We have thus got
    L = 4  M = 3 and R = 9

    $$\sqrt[3]{8\ 4\ 6\ 0\ 4\ 5\ 1\ 9} \qquad = 4\ 3\ 9$$

**Example 6:** Find the cube root of 279726264

**Solution:**

1.  Placing the bar over the number we find that there are three groups of numbers, so the cube root will have three digits.

    $$\sqrt[3]{2\ 7\ 9\ 7\ 2\ 6\ 2\ 6\ 4}$$

2.  From the Table 1, we can say that the unit digit of the cube root is 4. Hence R = 4.

3. Moreover, $6^3 < 33 < 7^3$ , so L = 6 (Refer Table 2)
4. Subtract $R^3$ from the number and cancel out the last zero.

```
  2 7 9 7 2 6 2 6 4
               − 6 4
  2 7 9 7 2 6 2 0 0̸
```

5. Middle digit of the cube root is obtained by 3 $R^2$ M. Here, 3 $R^2$ M. = 3 x $4^2$ x M = 48 M.
6. We are now looking for a suitable value of M so that the unit digit of 48 M becomes equal to the unit digit of the number obtained in step 4. Thus M = 5.
7. We have thus got
   L = 6   M = 5 and R = 4

   $$\sqrt[3]{2\,7\,9\,7\,2\,6\,2\,6\,4} = 6\,5\,4.$$

*Case 3: Cube Root of a number greater than 7 digits, but ending with even number.*

Many a time, while extracting the cube root of a number having its unit digit, even you may get two values of M and it becomes difficult to ascertain the exact value of M. In such a case, you may reach to the exact answer by two ways.

•   Divide the number whose cube root has to be extracted by 8, until an odd cube emanates
•   Use the Beejank (बीजांक) method to reach the exact value of M

Let us take some examples to understand this case.

**Example 7:** Find the cube root of 1906624

**Solution:** Since this is an even number, to avoid the ambiguous case in extracting the cube root, we need to divide the number by 8 until an odd cube emanates.

```
8 | 1 9 0 6 6 2 4
8 | 2 3 8 3 2 8
    2 9 7 9 1
```

The cube root of 29761 can be extracted merely by the Vilokanam (विलोकनम्) method using Table 1 and Table 2. You may refer to case1.

Group the number

$$\overline{2\ 9}\ \overline{7\ 9}\ 1$$

Since there are two groups, the cube root of 29791 contains a two digit number.

Unit digit = 1

Ten's digit = 3

Hence, the cube root of 29791 = 31

We have,

$$\sqrt[3]{1906624} = \sqrt[3]{8 \times 8 \times 29791}$$

$$= 2 \times 2 \times 31 = 124$$

**Example 8:** Find the cube root of 51478848?

**Solution:** Since this is an even number, to avoid the ambiguous case in extracting the cube root, we need to divide the number by 8 until an odd cube emanates.

```
8 | 51478848
8 | 6434856
    804357
```

The cube root of 804357 can be extracted merely by the Vilokanam (विलोकनम्) method using Table 1 and Table 2. You may refer to case 1.

Group the number

$$\overline{804}\ \overline{357}$$

Since there are two groups, the cube root of 804357 contains a two digit number.

Unit digit = 3 (Refer Table 1)

Since, $9^3 < 804 < 10^3$

Hence, Ten's digit = 9 (Refer Table 2)

We have,

$$\sqrt[3]{51478848} = \sqrt[3]{8 \times 8 \times 804357}$$

$$= 2 \times 2 \times 93 = 372$$

Let us extract the cube root by the ***Beejank method***.

Here the method is slightly changed. In case 2, we have subtracted $R^3$ from the number and neglected the last zero but here $L^3$ is subtracted. You may consider either of these cases.

1.  Group the number from right to left by placing the bar.

    $$\sqrt[3]{5\ 1\ 4\ 7\ 8\ 8\ 4\ 8}$$

2.  The unit digit of the cube root is 8 (Refer Table 1). Hence R = 2.

3.  More over $3^3 < 51 < 4^3$, so L = 3 (Refer Table 2)

4.  Subtract $L^3$ from the number.

    ```
      5 1 4 7 8 8 4 8
    - 2 7
    ─────────────────
      1 4 4
    ```

5.  Middle digit of the cube root is obtained by 3 $R^2$ M.

    3 $R^2$ M. = 3 x $2^2$ x M = 12 M

6.  Now, the bigger question is what value of M should be put in 12 M so that the unit digit ends with 4. We find here 2 options they are 2, and 7.

7.  This situation can be handled by applying the Beejank method

    Beejank of 51478848 = 0

    Beejank of $(322)^3$ = 1

    Beejank of $(372)^3$ = 0

This clearly shows that our answer is 372.

**Example 9:** Find the cube root of 7738893352

**Solution:** Since this is an even number, to avoid the ambiguous case in extracting the cube root, we need to divide the number by 8 until an odd cube emanates.

$$
\begin{array}{c|l}
8 & 7738893352 \\
\hline
 & 967361669 \\
\hline
\end{array}
$$

The cube root of 967361669 can be extracted by the **Beejank method**.

1. Group the number from right to left by placing the bar

   $\sqrt[3]{\overline{967}\ \overline{361}\ \overline{669}}$

2. The unit digit of the cube root is 9 (Refer Table 1). Hence R = 9

3. Moreover $9^3 < 967 < 10^3$, so L = 9 (Refer Table 2)

4. Subtract $R^3$ from the number and neglect the last zero.

   $$
   \begin{array}{r}
   9\ 6\ 7\ 3\ 6\ 1\ 6\ 6\ 9 \\
   -\ 7\ 2\ 9 \\
   \hline
   9\ 6\ 7\ 3\ 6\ 0\ 9\ 4\ \cancel{0} \\
   \end{array}
   $$

5. Middle digit of the cube root is obtained by 3 $R^2$ M.
   3 $R^2$ M. = 3 x $9^2$ x M = 243 M

6. Now, the bigger question is what value of M should be put in 243 M, so that the unit digit ends with 4.

This clearly shows that M = 8

Hence,

$$\sqrt[3]{\overline{967}\ \overline{361}\ \overline{669}} \qquad = 9\ 8\ 9$$

Therefore,

$$\sqrt[3]{7738893352} = \sqrt[3]{8 \times 967361669} \qquad = 2 \times 989 = 1978$$

# 12

# Fourth Power of a Number

## Introduction

In the previous chapter, we learnt to find the cube of a number. Likewise, we can find the fourth power of any two-digit number with the same technique and extend it for a higher power. Before we move to the Vedic method, I would like to introduce here a very special method called the Pascal Triangle that will be of immense use for students who want to learn the formula of binomials having any number of powers.

**Pascal Triangle:**

A triangular array of numbers is composed of the coefficients in the expansion of $(x + y)^n$ for n = 0, 1, 2, 3, .. etc. The triangle extends down infinitely, the coefficients in the expansion of $(x + y)^n$, being in the $(n + 1)^{th}$ row. The array described below is bordered by 1s and the sum of two adjacent numbers in one row is equal to the number in the next row between the two numbers. This array is symmetrical about the vertical line through the vertex.

| | | | | | | | | |
|---|---|---|---|---|---|---|---|---|
| N=1 | | | | | 1 | | | |
| N=2 | | | | 1 | | 2 | | 1 |
| N=3 | | | 1 | | 3 | | 3 | | 1 |
| N=4 | | 1 | | 4 | | 6 | | 4 | | 1 |
| N= 5 | 1 | | 5 | | 10 | | 10 | | 5 | | 1 |

Blaise Pascal was born on 19 Jun 1823 in France. He was a true genius. At the age of 12, he discovered that the sum of the angles of a triangle is 180 degrees. He invented the first digital calculator to help his father, who was a tax collector. He laid the foundation of the theory of Probability. His work on the topic in Tratise on the Arithmetical Traingle carried the most interesting triangle called the Pascal Triangle.

How does the Pascal Triangle method work?

**Rule:**

1. Look at the power of the binomial and move that many steps down. Note down the coefficients of that particular row and write it.
2. Write the maximum power of **a** and minimum power of **b** as first coefficient.
3. Decrease the power of **a** by 1 and increase the power of **b** by 1 in each of the coefficients staring from the left-most coefficient.
4. Continue the process until every coefficient is taken into account.

**Example:** Find $(a + b)^3$

**Solution:** Here, the power of the binomial expansion is 3 so move 3 step down and note down the coefficients from the above triangle. As you can see that the coefficients are – 1, 3, 3,and 1.

Assign the maximum power of a (i.e. 3)and minimum power to b (i.e. 0) in the first coefficient and keep decreasing the power of **a** by 1 and increasing the power of **b** by 1, until all coefficients are taken into account.

$$1\ a^3b^0 \qquad 3a^2b^1 \qquad 3a^1b^2 \qquad a^0\ b^3$$

Since we are expanding the binomial with (+) sign, put the (+) sign in between each assigned value in the above line. Replace the value of $a^0$ **and** $b^0$ by 1. Now the formula is–

$$(a + b)^3 = a^3 + \qquad 3a^2b^1 \quad + 3a^1b^2 + b^3$$

**Example 2**: Expand $(a + b)^4$

**Solution**: Here the power of the binomial is 4, so write down the coefficient of the fourth row. Assign the power as done above.

Coefficients are–    1      4      6      4      1

Assign the value of a and b

$$1\ a^4b^0 \qquad 4a^3b^1 \qquad 6a^2b^2 \qquad 4a^1b^3 \qquad 1a^0b^4$$

Place the (+) sign in between to get the formula.

$$
\begin{aligned}
(a+b)^4 &= 1\ a^4b^0 &&+ 4a^3b^1 &&+ 6a^2b^2 &&+ 4a^1b^3 &&+ 1a^0b^4 \\
&= a^4 &&+ 4a^3b &&+ 6a^2b^2 &&+ 4ab^3 &&+ b^4
\end{aligned}
$$

In case, there is a minus sign in between a and b place (+) and (−) sign alternatively.

**Example**: Expand $(a - b)^4$

**Solution**: Apply the above method and write down the coefficients and assign the value of a and b thereafter. Lastly, place the (+) and (−) sign alternatively.

$$
\begin{aligned}
(a+b)^4 &= 1\ a^4b^0 &&- 4a^3b^1 &&+ 6a^2b^2 &&- 4a^1b^3 &&+ 1a^0b^4 \\
&= a^4 &&- 4a^3b &&+ 6a^2b^2 &&- 4ab^3 &&+ b^4
\end{aligned}
$$

## Vedic Method:

From the above explanation, you must have learnt to find the expansion of any binomial. The above Pascal Triangle is for n =5, but you can extend it for more values of n.

Now, let's return to our business. The above expansion of $(a+b)^4$ can be re-written as-

$$
\begin{aligned}
(a+b)^4 \quad &= \quad a^4 \;+\; a^3b \;+\; a^2b^2 \;+\; ab^3 \;+\; b^4 \\
& \qquad\qquad\; 3\ a^3b \;+\; 5a^2b^2 \;+\; 3ab^3
\end{aligned}
$$

Hence, $(a+b)^4 = a^4 + 4a^3b + 6a^2b^2 + 4ab^3 + b^4$

If we consider,

a = first term (Digit at ten's place)

b = second term (Digit at unit place)

Hence, ratio = b/a

Looking upon the above explanation, we can notice that subsequent to the first term, each of the remaining terms is obtained by multiplying the previous term by the ratio (b/a).

**Rule:**

- If you are to find the fourth power of any two digit number, first find the ratio as explained above. Ratio = unit digit/ ten's digit
- Raise the fourth power to the first term (i.e., the digit at ten's place), call it a.
- Multiply each subsequent term by the ratio and obtain the remaining 4 terms.

  | $a^4$ | $a^4$ x b/a | $a^3b$ x b/a | $a^2b^2$ x b/a | $ab^3$ x b/a |
  |-------|-------------|--------------|----------------|--------------|
  | $a^4$ | $a^3b$      | $a^2b^2$     | $ab^3$         | $b^4$        |

- Multiply the second term by 3, the third term with 5 and fourth term with 3 and write it in the second row.
- Add each column (from right to left)
- No column (except the first) should contain more than 1 digit. If the sum of any column has more than 1 digit, transfer it to the preceding column.

**Example:** Find the fourth power of 12.

**Solution:** In $(12)^4$, a = 1 and b = 2. Hence, ratio = b/a = 2/1 = 2

- Make a box having 5 columns.

| 1$^{st}$ column | 2$^{nd}$ column | 3$^{rd}$ column | 4$^{th}$ column | 5$^{th}$ column |
|-----------------|-----------------|-----------------|-----------------|-----------------|
|                 |                 |                 |                 |                 |

- Raise fourth power of **a** and place it in 1$^{st}$ column. Fill the rest column by multiplying the digit in the first column with ratio 2.

| 1st column | 2nd column | 3rd column | 4th column | 5th column |
|---|---|---|---|---|
| $1^4 = 1$ | 1 x 2 = 2 | 2 x 2 = 4 | 4 x 2 = 8 | 8 x 2 = 16 |

- Multiply the digit at second column with 3, the third column digit with 5 and fourth column digit with 3 and place it in 2nd row of the table.

| 1st column | 2nd column | 3rd column | 4th column | 5th column |
|---|---|---|---|---|
| 1 | 2 | 4 | 8 | 16 |
|  | 2 x 3 = 6 | 4 x 5 = 20 | 8 x 3 = 24 |  |

- Add each column separately.

| 1st column | 2nd column | 3rd column | 4th column | 5th column |
|---|---|---|---|---|
| 1 | 2 | 4 | 8 | 16 |
|  | 6 | 20 | 24 |  |
| 1 | 8 | 24 | 32 | 16 |

- Transfer the exceeding digit from each column to its preceding column.

| 1st column | 2nd column | 3rd column | 4th column | 5th column |
|---|---|---|---|---|
| 1 | 2 | 4 | 8 | 16 |
|  | 6 | 20 | 24 |  |
| 1 | 8 | 24 | 32 | 16 |

| 1 | 8 | 24 | 33(32 +1) | 6 |
|---|---|---|---|---|

| 1 | 8 | 27(24 +3) | 3 | 6 |
|---|---|---|---|---|

| 1 | 10(8 +2) | 7 | 3 | 6 |
|---|---|---|---|---|

| 2 (1 +1) | 0 | 7 | 3 | 6 |
|---|---|---|---|---|

Hence, $(12)^4 = 20736$

**Example 2:** Find the fourth power of 21

**Solution:** Here, in $(21)^4$, a = 2 and b = 1 and ratio = b/a = 1/ 2
Make a box having 5 columns.

| 1st column | 2nd column | 3rd column | 4th column | 5th column |
|---|---|---|---|---|
|  |  |  |  |  |

- Raise fourth power of **a** and place it in 1st column. Fill the rest column by multiplying the digit in the first column with ratio 2.

| 1st column | 2nd column | 3rd column | 4th column | 5th column |
|---|---|---|---|---|
| $2^4$ = 16 | 16 x 1/2 = 8 | 8 x 1/ 2 = 4 | 4 x 1/2 = 2 | 2 x 1/ 2 = 1 |

- Multiply the digit in the second column with 3, the third column digit with 5 and fourth column digit with 3 and place it in the 2nd row of the table.

| 1st column | 2nd column | 3rd column | 4th column | 5th column |
|---|---|---|---|---|
| 16 | 8 | 4 | 2 | 1 |
|  | 8 x 3 =24 | 4 x 5 =20 | 2 x 3 = 6 |  |

- Add each column separately.

| 1st column | 2nd column | 3rd column | 4th column | 5th column |
|---|---|---|---|---|
| 16 | 8 | 4 | 2 | 1 |
|  | 24 | 20 | 6 |  |
| 16 | 32 | 24 | 8 | 1 |

- Transfer the exceeding digit from each column to its preceding column.

| 1st column | 2nd column | 3rd column | 4th column | 5th column |
|---|---|---|---|---|
| 16 | 8 | 4 | 2 | 1 |
|  | 24 | 20 | 6 |  |

| 16 | 32 | 24 | 8 | 1 |
|---|---|---|---|---|

| 16 | 34(32+2) | 4 | 8 | 1 |
|---|---|---|---|---|

| 19(16+3) | 4 | 4 | 8 | 1 |
|---|---|---|---|---|

Hence $(21)^4 = 194481$

**Example 3:** Find the fourth power of 11

**Solution:** In $(11)^4$, a = 1 and b = 1. Here ratio = b/a = 1

Write the fourth power of a in the $1^{st}$ column and the remaining 4 terms in the rest of the columns as done in the previous two example. The table will look like

| $1^{st}$ column | $2^{nd}$ column | $3^{rd}$ column | $4^{th}$ column | $5^{th}$ column |
|---|---|---|---|---|
| $1^4 = 1$ | 1 x 1 = 1 | 1 x 1 = 1 | 1 x 1 = 1 | 1 x 1 = 1 |

- Multiply the digit at second column with 3, the third column digit with 5 and fourth column digit with 3 and place it in $2^{nd}$ row of the table.

| $1^{st}$ column | $2^{nd}$ column | $3^{rd}$ column | $4^{th}$ column | $5^{th}$ column |
|---|---|---|---|---|
| 1 | 1 | 1 | 1 | 1 |
|  | 1 x3 =3 | 1 x 5 = 5 | 1 x 3 = 3 |  |

- Add each column separately.

| $1^{st}$ column | $2^{nd}$ column | $3^{rd}$ column | $4^{th}$ column | $5^{th}$ column |
|---|---|---|---|---|
| 1 | 1 | 1 | 1 | 1 |
|  | 3 | 5 | 3 |  |
| 1 | 4 | 6 | 4 | 1 |

Since, each column contains only single digit therefore this is the answer.

**Example 4:** Find the fourth power of 14?

**Solution:** Here , a = 1 and b = 4 in $(14)^4$ . Ratio = 4/1 = 4

Make a table of 5 column and place the value in each column accordingly. Multiply the digit at second column with 3, the third column digit with the 5 and fourth column digit with 3 and place it in $2^{nd}$ row of the table. The table will now look as–

| $1^{st}$ column | $2^{nd}$ column | $3^{rd}$ column | $4^{th}$ column | $5^{th}$ column |
|---|---|---|---|---|
| 1 | 4 | 16 | 64 | 256 |
| | 4 x 3 = 12 | 16 x 5 = 80 | 64 x 3 = 192 | |

- Add each column separately

| $1^{st}$ column | $2^{nd}$ column | $3^{rd}$ column | $4^{th}$ column | $5^{th}$ column |
|---|---|---|---|---|
| 1 | 4 | 16 | 64 | 256 |
| | 4 x3 =12 | 16 x 5 = 80 | 64 x 3 = 192 | |
| 1 | 16 | 96 | 256 | 256 |

- Transfer the exceeding digit from each column to its preceding column.

| $1^{st}$ column | $2^{nd}$ column | $3^{rd}$ column | $4^{th}$ column | $5^{th}$ column |
|---|---|---|---|---|
| 1 | 4 | 16 | 64 | 256 |
| | 4 x 3 =12 | 16 x 5 = 80 | 64 x 3 = 192 | |
| 1 | 16 | 96 | 256 | 256 |

| $1^{st}$ column | $2^{nd}$ column | $3^{rd}$ column | $4^{th}$ column | $5^{th}$ column |
|---|---|---|---|---|
| 1 | 4 | 16 | 64 | 256 |
| | 12 | 80 | 192 | |

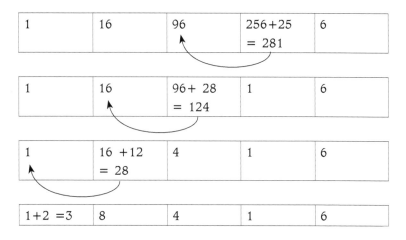

| 1 | 16 | 96 | 256+25 = 281 | 6 |

| 1 | 16 | 96+ 28 = 124 | 1 | 6 |

| 1 | 16 +12 = 28 | 4 | 1 | 6 |

| 1+2 =3 | 8 | 4 | 1 | 6 |

Hence, $(14)^4 = 38416$

**Example:** Find the fourth power of 45?

**Solution:** Here, a = 4 and b = 5 in $(45)^4$. Ratio = 5/4

• Make a table of 5 column and place the value in each column accordingly. Multiply the digit at second column with 3, the third column digit with 5 and fourth column digit with 3 and place it in $2^{nd}$ row of the table. The table will now look as–

| $1^{st}$ column | $2^{nd}$ column | $3^{rd}$ column | $4^{th}$ column | $5^{th}$ column |
|---|---|---|---|---|
| 256 | 320 | 400 | 500 | 625 |
|  | 960 | 2000 | 1500 |  |

• Add each column separately

| $1^{st}$ column | $2^{nd}$ column | $3^{rd}$ column | $4^{th}$ column | $5^{th}$ column |
|---|---|---|---|---|
| 256 | 320 | 400 | 500 | 625 |
|  | 960 | 2000 | 1500 |  |
| 256 | 1280 | 2400 | 2000 | 625 |

- Transfer the exceeding digit from each column to its preceding column.

| 1$^{st}$ column | 2$^{nd}$ column | 3$^{rd}$ column | 4$^{th}$ column | 5$^{th}$ column |
|---|---|---|---|---|
| 256 | 320 | 400 | 500 | 625 |
|  | 960 | 2000 | 1500 |  |
| 256 | 1280 | 2400 | 2000 | 625 |

| 256 | 1280 | 2400 | 2062 | 5 |
|---|---|---|---|---|

| 256 | 1280 | 2606 | 2 | 5 |
|---|---|---|---|---|

| 256 | 1540 | 6 | 2 | 5 |
|---|---|---|---|---|

| 410 | 0 | 6 | 2 | 5 |
|---|---|---|---|---|

Hence, $(45)^4 = 4100625$

I do hope the above method with a little practice will reduce calculation time by 1/10th of the actual time taken in calculating the same by traditional method.

# 13

# Fourth Root of a Number

## Introduction

If, $x^4 = 16$ so, $x = \sqrt[4]{16}$

$$= \sqrt[4]{2 \times 2 \times 2 \times 2}$$
$$= 2$$

The above example clearly shows that in order to get the fourth power of a desired number, we break it into 4 equal factors and take one factor out of four.

In general, there is no traditional method taught in our curriculum that helps us to find the fourth root of a number, except the labyrinth Prime Factor Method. The traditional method is too clumsy and time-consuming. As long as the number is 2–4 digits, the prime factor method works effectively, but if the number is more than 4 digits, this method takes more than 5 minutes to arrive at an answer.

Let us understand it with an example;-

**Example:** Find the fourth power of 1679616
Solution: The calculation in the traditional method looks like-

| 4 | 1679616 |
|---|---|
| 4 | 419904 |
| 4 | 104976 |
| 4 | 26244 |
| 3 | 6561 |
| 3 | 2187 |
| 3 | 729 |
| 3 | 243 |
| 3 | 81 |
| 3 | 27 |
| 3 | 9 |
| 3 | 3 |
|   | 1 |

Hence, $\sqrt[4]{1679616} = \sqrt[4]{4 \times 4 \times 4 \times 4 \ \times \ 3 \times 3 \times 3 \times 3 \ \times \ 3 \times 3 \times 3 \times 3}$

$$= 4 \times 3 \times 3 = 36$$

The total time taken in prime factorization and to extract the fourth root is more than 5 minutes and it is not the right process that you can rely upon. Don't worry – here is an easy and fantastic method to reduce your timing from 5 minutes to 5 seconds.

## Vedic Sutra for extracting fourth power
## Vilokanam (विलोकनम्)

**Meaning of Vedic Sutra:** The word Vilokanam means, "by mere observation". The first table will help you to find the unit digit of the exact square root, whereas Table 2 will help you to find the ten's digit.

Let us first look at the following table:

### Table 1

| Digit | Fourth power | Single sum digit |
|-------|--------------|------------------|
| 1 | 1 | 1 |
| 2 | 16 | 7 |
| 3 | 81 | 9 |
| 4 | 256 | 4 |
| 5 | 625 | 4 |
| 6 | 1296 | 9 |
| 7 | 2401 | 7 |
| 8 | 4096 | 1 |
| 9 | 6561 | 9 |
| 10 | 10000 | 1 |

From the above table, we conclude:-
- Any number whose unit digit is 1, will have 1, 3, 7 ,or 9 at the unit digit of the fourth power
- Any number whose unit digit is 6, will have 2, 4, 6, or 8 at the unit digit of the fourth power.
- Any number whose unit digit is 5, will have 5 at the unit digit of the fourth power.
- If a number ends with 0000, it will have 0 at the unit digit in the fourth power.

Let us summarise the above facts in a table.

### Table 2

| Number | Nearest fourth Root |
|--------|---------------------|
| 1 — 15 | 1 |
| 16 — 80 | 2 |
| 81 — 255 | 3 |
| 256 — 624 | 4 |

| | |
|---|---|
| 625 – 1295 | 5 |
| 1296 – 2400 | 6 |
| 2401 – 4095 | 7 |
| 4096 – 6560 | 8 |
| 6561 – 9999 | 9 |

## Fundamental rule for extracting the fourth root of a number

- Make groups of 4 digits from the right.
- The $1^{st}$ pair containing the unit digit will give you the unit place of the fourth root of a number (see Table 1). Let us denote the unit pair by A.
- Look at the value of the second pair in the Table 2 to identify the ten's digit of the fourth root of a number. Let us call the ten's digit of the fourth root by B.
- If you are having a different set of options, subtract $B^4$ from the 2nd pair and carry down the next digit from the number as done in simple division.
- Now compare the new dividend with $4 B^3 A$. Put different values of A and select the best possible digit, so that the product $4 B^3 A \leq$ the new dividend
- These two digits together give the fourth root of a number.

**Example 1:** Find the fourth root of 6765201

**Solution:**

- Make groups of 4 digits from the right.

$$\underset{2^{nd} \text{ Pair (B)}}{\underline{676}} \quad \underset{1^{st} \text{ Pair (A)}}{\underline{5201}}$$

- Since the unit digit of the number is 1, possible digits at the unit place of the fourth root are 1, 3, 7, or 9 (See Table 1).

- Look at the value of the 2$^{nd}$ pair in Table 2 to determine the value of B. Here 625 < 676 < 1295, so B = 5.
- Hence, the possible set of answers are – 51, 53, 57 or 59.
- Subtract B$^4$ from the given number and carry down the next digit from the number to get the new dividend.

676    5 2 0 1

$-5^4$    $\downarrow$ ———————➤ Carry down the next digit from the number

51    5 ———————➤ new dividend

- Compare the new dividend by 4 B$^3$A
- We have A = 1 , 3, 7, or 9 and B = 5

| A | B | 4 B$^3$ A |
|---|---|---|
| 1 | 5 | 500 |
| 3 | 5 | 1500 |
| 7 | 5 | 3500 |
| 9 | 5 | 4500 |

- Since our new dividend = 515 and the above table shows that only A = 1 satisfies the result, we reject the other option and take A = 1. This might seem lengthy to you, but with a little practice, you will be able to guess the result quite comfortably, without setting different values of A. From the above table, it is clear that you could have saved time without calculating the value of A = 3, 5 or 7.
- Hence, the fourth root of 6765201 = 51

**Example 2:** Find the fourth root of 25411681

**Solution:**

Make groups of 4 digits from the right.

2541    1681

2$^{nd}$ Pair (B)⤢          ⤢ 1$^{st}$ Pair (A)

- Since the unit digit of the number is 1, possible digits

at the unit place of the fourth root are 1, 3, 7, or 9 (See Table 1).

- Look at the value of 2$^{nd}$ pair in the Table 2 to determine the value of B. Here 2401 < 2541 < 4095 so B = 7.
- Hence, the possible set of answers are – 71, 73, 77 or 79.
- Subtract B$^4$ from the given number and carry down the next digit from the number to get the new dividend.

$$2541 \quad 1 \; 6 \; 8 \; 1$$
$$\underline{-7^4} \qquad \longrightarrow \text{Carry down the next digit from the number}$$
$$140 \quad 1 \longrightarrow \text{new dividend}$$

- Compare the new dividend by 4 B$^3$A
- We have, A = 1 , 3, 7, or 9 and B = 7

| A | B | 4 B$^3$ | 4 B$^3$ A |
|---|---|---|---|
| 1 | 7 | 1372 | 1372 |
| 3 | 7 | ------------- | ----------- |
| 7 | 7 | ----------- | ------------ |
| 9 | 7 | ----------- | ----------- |

- Since our new dividend = 1401 and the above table shows that only A = 1 satisfies the result, we reject the other option and take A = 1. As 1372 x 3 > 1401, we stop taking other options for A. Hence, the fourth root of 25411681 = 71

**Example 3:** Find the fourth root of 5308416

**Solution:**

- Make groups of 4 digits from the right.

$$\underline{530} \quad \underline{8416}$$
$$\text{2}^{nd} \text{ Pair (B)} \qquad \text{1}^{st} \text{ Pair (A)}$$

- Since the unit digit of the number is 6, possible digits at the unit place of the fourth root are 2, 4, 6, or 8 (See Table 1).
- Look at the value of the $2^{nd}$ pair in the Table 2 to determine the value of B. Here $256 < 530 < 624$ so B = 4.
- Hence, the possible sets of answers are– 42, 44, 46, or 48.
- Subtract $B^4$ from the given number and carry down the next digit from the number to get the new dividend.

$$\begin{array}{ll} 530 & 8\ 4\ 1\ 6 \\ -4^4 & \\ \hline 274 & 8 \end{array}$$

→ Carry down the next digit from the number

→ new dividend

- Compare the new dividend by 4 $B^3$A
- We have A = 2 , 4, 6, or 8 and B = 4

| A | B | 4 $B^3$ | 4 $B^3$ A |
|---|---|---|---|
| 2 | 4 | 256 | 512 |
| 4 | 4 | 256 | 1024 |
| 6 | 4 | 256 | 1536 |
| 8 | 4 | 256 | 2048 |

- Since our new dividend = 2748 and the above table shows that only A = 8 satisfies the result, the fourth root of 53088416 = 48.

# 14

# Simultaneous Equation

## Introduction

Two linear equations in two unknown x and y are said to form a system of simultaneous equations if each of them is satisfied by the same pair of value of x and y.

**Example:** $x + y = 4$ and $x - y = 3$ is an example of simultaneous equation.

In our school curriculum, there are various methods taught to solve the linear simultaneous equation, but none of them are time saving.

- Method of Elimination
- Method of Comparison
- Method of Substitution
- Method of Cross Multiplication

The cross multiplication method is fast, but its main drawback is its confusing nature; and the important fact is that even teachers get confused regarding the plus or minus sign. The Vedic method on the other hand, is fast and simple. It makes calculation easy and far-quicker than the traditional method. The most important feature of Vedic Mathematics is its various techniques to solve the problem in no time, using easy and convenient methods.

## Vedic Sutras for solving Simultaneous Equation

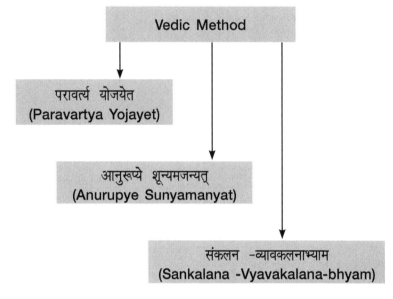

**Meaning of Vedic Sutra**

1. **Paravartya Yojayet** (परावर्त्य योजयेत्): The literal meaning of this sutra is – Transpose and Apply. The well known rule relating to transposition enjoins an invariable change of sign with every change in side.

2. **Anurupye Sunyamanyat** (अनूरूप्ये शून्यमन्यत्): The sutra says – If one is in ratio, the other one is zero. The detailed explanation with examples is given below.

3. **Sankalana – Vyavakalana- bhyam** (संकलन व्यवकलनाभ्याम्): This sutra has also been used in the chapter on Subtraction. The simple meaning is Addition and Subtraction.

## Paravartya Yojayet

This method is applicable for all sorts of linear simultaneous equations. The cross multiplication method taught in our present day curriculum is somewhat akin to the Paravartya sutra. The Vedic Sutra moves in a cyclic order.

For x, we start with the y coefficient and the independent terms and cross multiply them in the forward direction. The sign between the two cross multiplications is minus $(-)$.

For y, we start with the independent term and x coefficients and cross multiply them in the backward direction. The sign between the cross multiplication result is minus $(-)$.

For the result of the denominator, we take the coefficient of variables only and cross multiply them in backward direction.

Suppose we have the following set of simultaneous equations:-

$$a_1x + b_1y = c_1$$
$$a_2x + b_2y = c_2$$

In order to get the numerator of x, we leave the coefficients of x and write the coefficient of y and the independent term and cross multiply them with a minus sign in between the cross product in the rightward direction as shown here.

$$\begin{matrix} b_1 & c_{1\ (2)} \\ b_2 & c_{2\ (1)} \end{matrix}$$
$$= b_1\,c_2 - b_2\,c_1$$

Again, to get the numerator of y, we leave the coefficient of y and take only the coefficient of x and the independent term into consideration. As you know, the sutra moves in a cyclic order, so we have to start with the independent term first. Cross multiplication of the independent term and coefficient of x will give the numerator of y.

$$\begin{matrix} c_1 & a_{1\ (1)} \\ c_2 & a_{2\ (2)} \end{matrix}$$
$$c_1\,a_2 - a_1\,c_2$$

The denominator of both the variables x and y will remain same. The cross product of the coefficient of the variable in the backward direction gives us the result:

$$\begin{matrix} a_1 & b_{1\ (1)} \\ a_2 & b_{2\ (2)} \end{matrix}$$

To make the concept more clear let us put the whole thing in a simple diagrammatical structure.

Nr of x          Nr of y          Denominator (For x and y)

$$b_1 \quad c_1 \quad a_1 \quad b_1$$
$$b_2 \quad c_2 \quad a_2 \quad b_2$$

Let us take a few examples to understand the modus operandi. Here Nr defines Numerator.

**Example** 1: Solve for x and y: $2x + 3y = 7$ ; $3x + 7y = 13$.

**Solution**:

Coefficient of y          Independent Term

3        7

$x =$    7        13

Coefficient of x          Coefficient of y

2        3

3        7

Independent Term          Coefficient of x

7        2

$y =$    13        3

Coefficient of x          Coefficient of y

2        3

3        7

Hence, $x = \dfrac{39 - 49}{9 - 14} = \dfrac{-10}{-5} = 2$

$y = \dfrac{21 - 26}{9 - 14} = \dfrac{-5}{-5} = 1$

**Example** 1: Solve for x and y: 11 x + 6y = 21 ; 8x −5y = 34.

**Solution:**

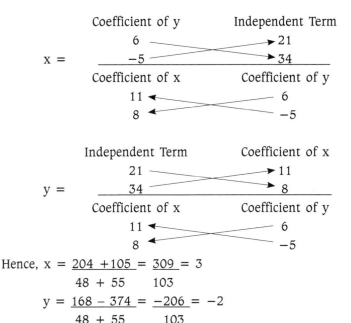

Hence, x = $\dfrac{204 + 105}{48 + 55}$ = $\dfrac{309}{103}$ = 3

y = $\dfrac{168 - 374}{48 + 55}$ = $\dfrac{-206}{103}$ = −2

**Example** 1: Solve for x and y: 4x + 7y = 29 ; 12x + 3y = −3.

**Solution:**

|  | Independent Term | Coefficient of x |
|---|---|---|

$$y = \frac{\begin{array}{cc} 29 & 4 \\ -3 & 12 \end{array}}{\begin{array}{cc} 4 & 7 \\ 12 & 3 \end{array}}$$

Independent Term     Coefficient of x

29       4

−3       12

Coefficient of x     Coefficient of y

4       7

12       3

Hence, $x = \dfrac{-21 - 87}{84 - 12} = \dfrac{-108}{-72} = \dfrac{-3}{2}$

$y = \dfrac{348 + 12}{84 - 12} = \dfrac{360}{72} = 5$

## Anurupye Sunyamanyat (अनुरूप्ये सून्यमन्यत्)

This is a special type of simultaneous equation, which at the very first instance looks very hard due to the involvement of large numbers, but the Vedic Method makes it child's play. The Vedic Sutra says – If one is in ratio, the other one is zero. In simple language, whenever the ratio of x or y is equal to that of the independent term, put the ratio of y or x = 0.

Let me take one example to make the modus operandi understandable to you.

**Example**: Solve for x and y:

$$5x + 8y = 40$$
$$10x + 11y = 80$$

**Solution**: In the above example, the ratio of the coefficients of x is 1: 2 and the ratio of the independent term is also 1: 2. The Vedic sutra in this special case says – If one is in ratio, the other one is zero.

Since the ratio of x is equal to the ratio of the independent term, y = 0. Put y = 0 in either of the equations to get the value of x.

For y = 0, 10 x = 80 hence x = 8

x = 8 and y = 0 is the solution.

**Example:** Solve for x and y:

$$12 x + 78 y = 12$$
$$16 x + 96 y = 16$$

**Solution:** Here the ratio of x = 12: 16 = ratio of independent term

Hence, y = 0
Put y = 0 in either of the two equations to get x = 1

**Example:** Solve for x and y:

$$44x + 178 y = 22$$
$$132 x + 243 y = 66$$

**Solution:** Here the ratio of coefficient of x = 44: 132 = 1: 3

The ratio of independent term = 22: 66 = 1: 3
Hence, y = 0
Put y = 0 to get x = ½.

**Example:** Solve for x and y:

$$27 x + 144 y = 720$$
$$42 x + 72 y = 360$$

**Solution:** Here the ratio of coefficient of y = 144: 72 = 2: 1

The ratio of independent term = 720: 360 =2:1
Hence, x = 0
Put x = 0 in any of the above equation to get y = 5.

## Sankalana Vyavakalanam (संकलन् व्यवकलनाभ्याम्)

As said earlier, this sutra simply means – Addition and Subtraction. Whenever the coefficient of x in the 1ˢᵗ equation is equal to the coefficient of y in the 2ⁿᵈ equation and vice versa, this sutra works better.

Let us take an example and see its modus operandi.

**Example**: Solve for x and y

$$23x + 31y = 18$$
$$31x + 23y = 90$$

**Solution**: In the above example

The coefficient of x in $1^{st}$ equation = Coefficient of y in $2^{nd}$ equation = 23
The coefficient of y in $1^{st}$ equation = coefficient of x in $2^{nd}$ equation = 31

Let us apply the Sankalana Vyavkalanam Vedic sutra to get a quick answer.
Adding the two equations we get,
$54x + 54y = 108 \Rightarrow x + y = 2$      ------(A)
On subtracting we get,
$-8x + 8y = -72, \Rightarrow -x + y = -9$      -------(B)
Add equation (A) and (B) to get $y = -7/2$
On subtracting equations (A) and (B) again we get,
$x = 1/2$

**Example**: Solve for x and y:

$$45x - 23y = 113$$
$$23x - 45y = 91$$

**Solution**: We have,

$$45x - 23y = 113 \quad ----- (1)$$
$$23x - 45y = 91 \quad ------(2)$$

Adding (1) and (2) we get,
$68x - 68y = 204 \Rightarrow x - y = 3$      -------(3)
Subtracting (2) from (1) we get,
$22x + 22y = 22 \Rightarrow x + y = 1$      -------(4)
Again on adding equation (3) and (4) we get, $x = 2$
And on subtracting (4) from (3) we get, $y = -1$

**Example**: Solve for x and y :

$$699 x + 845 y = 5477$$
$$845 x + 699 y = 5331$$

**Solution**: We have,

$$699 x + 845 y = 5477 \qquad ----(1)$$
$$845 x + 699 y = 5331 \qquad ----(2)$$

Add (1) and (2)

$$\Rightarrow 1544 x + 1544 y = 10808 \Rightarrow x + y = 7 \quad --(3)$$

Subtract (2) from (1) we get,

$$- 146 x + 146 y = 146 \Rightarrow -x + y = 1 \qquad --(4)$$

Add (3) and (4) we get, y = 4

Subtract (4) from (3) we get, x = 3

**Example**: Solve for x and y:

$$23 x - 29 y = 98$$
$$29 x - 23 y = 110$$

**Solution**: We have,

$$23 x - 29 y = 98 ---(1)$$
$$29 x - 23 y = 110 --(2)$$

Adding (1) and (2) we get,

$$52 x - 52 y = 208 \Rightarrow x - y = 4 \qquad ---- (3)$$

Subtracting (2) from (1) we have,

$$- 6 x - 6 y = - 12 \Rightarrow x + y = 2 \qquad ---- (4)$$

Add (3) and (4) again to get x = 3

Subtract (4) from (3) and get, y = 1

I do hope the journey of linear simultaneous equation with Vedic sutras has been entertaining and solving these equations with the help of Vedic sutra will undoubtedly save your precious time and energy.

# Cubic Factorization

## Introduction

In NCERT text books of class 9, students encounter Cubic Factorization and do solve it by means of the Factor Theorem. The factor theorem method is easy, but is a time-consuming process. For some of you who are not aware of the factor theorem, let me first describe its definition and working.

Factor Theorem: If *p(x)* is a polynomial, then if *p(r) = 0,* then *(x – r)* is a factor of *p(x).*

Let us take an example to understand the tediousness of the Factor theorem.

**Example:** Factorize $x^3 + 6x^2 + 11x + 6$.

**Solution:**

Let $P(x) = x^3 + 6x^2 + 11x + 6$.

Since 6 has the factor $\pm 1$, $\pm 2$, $\pm 3$, and $\pm 6$, we invariably put these values in the polynomial p(x) until we get the remainder 0. Let us see how it works.

$P(x) = x^3 + 6x^2 + 11x + 6$.
$P(1) = (1)^3 + 6.(1)^2 + 11(1) + 6 \neq 0$
$P(-1) = (-1)^3 + 6.(-1)^2 + 11(-1) + 6$
$\qquad = -1 + 6 - 11 + 6 = 0$

Hence x + 1 is a factor of the polynomial P(x).
Now we divide P(x) by x +1 by long division method.

$$x + 1 \overline{\smash{\big)}\ x^3 + 6x^2 + 11x + 6}\ \big(\ x^2 + 5x + 6$$

$$\underline{x^3 + x^2}$$

$$5x^2 + 11x$$
$$\underline{5x^2 + 5x}$$

$$6x + 6$$
$$\underline{6x + 6}$$

Hence, $P(x) = x^3 + 6x^2 + 11x + 6 = (x + 1)(x^2 + 5x + 6)$

Now let us factorize the polynomials $x^2 + 5x + 6$ by splitting the middle term.

$$
\begin{aligned}
Q(x) &= x^2 + 5x + 6 \\
&= x^2 + 2x + 3x + 6 \\
&= x(x + 2) + 3(x + 2) \\
&= (x + 2)(x + 3).
\end{aligned}
$$

$$
\begin{aligned}
\text{So, } P(x) &= x^3 + 6x^2 + 11x + 6 \\
&= (x + 1)(x^2 + 5x + 6) \\
&= (x+1)(x + 2)(x + 3)
\end{aligned}
$$

Though the discussed method is easy but isn't it a labyrinth-like method to use?

## Formation of Cubic Polynomial

**Standard form:** $ax^3 + bx^2 + cx + d$

By the fundamental theorem of Algebra - Every polynomial of n degree has n roots. This is a cubic polynomial, so it will have three roots. Let us understand the structure of cubic polynomials.

Let the three roots be namely $\alpha$, $\beta$ and $\gamma$.

$\alpha + \beta + \gamma = $ sum of roots $= \Sigma\alpha = -b / a$

$\alpha\beta + \beta\gamma + \alpha\gamma$ = sum of product of two = $\Sigma\alpha\beta$ = c / a
$\alpha\beta\gamma$ = Product of three roots = $-d$ / a

Now let us see how Vedic Mathematics is handy in solving the cubic polynomials.

## Vedic Method

We first begin with the sub-sutra- Gunita Samuccaye Samuccaye Gunita (गुणितः समुच्चयः समुच्चय गुणितः) which means –
The product of the sum of the coefficients in the factors is equal to the sum of the coefficients in the product.
In symbols,

$S_e$ of Product = Product of $S_e$ (in the factor)

This sub-sutra is of immense utility for the purpose of verifying the correctness of our answer.

**Example:** $P(x) = x^3 + ax^2 + bx + c$

**Solution:** If the polynomial is cubic, it will have three roots - call it be $\alpha$, $\beta$ and $\gamma$.

Hence,
$a = \alpha + \beta + \gamma$ = sum of roots = coefficient of $x^2$
$b = \alpha\beta + \beta\gamma + \alpha\gamma$ = sum of the product of two roots = coefficient of x
$c = \alpha\beta\gamma$ = Product of three roots

Let us take a few examples to understand the concept of the Vedic Method to solve the factorization of cubic polynomials. You will be astonished to know that the Vedic method takes just 3–4 seconds to reach the answer, whereas the traditional methods taught in the classroom take at least 10 minutes. Moreover, there is little or no chance of faulty calculation. The present day textbook, though, uses Factor theorem and you can't use the Vedic method in your classroom as this method is not applicable in the traditional way, but at least you can use it to check your answer. Within a few seconds you can know the exact factor and thereafter can put only those values of x and apply the Factor Theorem method to reach the answer in half the time you presently devote.

**Example:** Factorize $x^3 + 6x^2 + 11x + 6$

**Solution:** The last term 6 has the factors 1, 2, 3 and 6. Our aim is to find $\alpha$, $\beta$ and $\gamma$ in such a way that it satisfies the following two conditions.

$c = \alpha\beta\gamma$ = Product of three roots = $1 \times 2 \times 3$
$a = \alpha+\beta + \gamma$ = sum of roots = coefficient of $x^2$ = $1 + 2 + 3$
Hence the factor is --- $(x + 1) (x + 2) (x + 3)$

**Verification:**

Put the value of $\alpha$, $\beta$ and $\gamma$ in b and check whether the coefficient of x in the polynomial is the same as the result you obtained or not.

$b = \alpha\beta + \beta\gamma + \alpha\gamma$ = sum of product of two = coefficient of x
$= 1 \times 2 + 2 \times 3 + 1 \times 3$
$= 2 + 6 + 3$
$= 11$
It is obvious from the result that the factor is absolutely correct.

**Example:** Factorize $x^3 + 12x^2 + 44x + 48$

**Solution:** Here the factor of 48 are 1, 2, 3, 4, 6, 8, 12, 16, 24 and 48.

Our aim is to choose three factors whose product is 48 and their sum is 12. In other words -

$a = \alpha+\beta + \gamma$ = sum of roots = coefficient of $x^2$ = 12
$c = \alpha\beta\gamma$ = Product of three roots = 48
On mere inspection, we see–
$\alpha\beta\gamma$ = Product of three roots = $2 \times 4 \times 6 = 48$

And $\alpha + \beta + \gamma$ = sum of roots = 2 + 4 + 6 = 12
Hence, the factors are–

$$x^3 + 12x^2 + 44x + 48 = (x + 2) \ (x + 4) \ (x + 6)$$

**Verification:**

$b = \alpha\beta + \beta\gamma + \alpha\gamma$ = sum of the product of two = coefficient of x
   = 2 x 4 + 4 x 6 + 2 x 6
   = 8 + 24 + 12
   = 44

Since, it is equal to the coefficient of x in the given polynomial, the factor thus obtained is correct.

**Example:** Factorize $x^3 + 8 \ x^2 + 19 \ x + 12$

**Solution:** The factors of 12 are– 1, 2, 3, 4, 6and 12.

$a = \alpha + \beta + \gamma$ = sum of roots = coefficient of $x^2$ = 8
$c = \alpha\beta\gamma$ = Product of three roots = 12
On mere inspection, we see -
$\alpha\beta\gamma$ = Product of three roots = 1 x 3 x 4 = 12
And $\alpha + \beta + \gamma$ = sum of roots = 1 +3 + 4 = 8
Hence, the factors are-

$$x^3 + 8 \ x^2 + 19 \ x + 12 = (x + 1) \ (x +3) \ (x + 4)$$

**Verification:**

$b = \alpha\beta + \beta\gamma + \alpha\gamma$ = sum of the product of two = coefficient of x
   = 1 x 3 + 3 x 4 + 1 x 4
   = 3 + 12 + 4
   = 19

**Example:** Factorize $x^3 - 2 \ x^2 - 33 \ x + 60$

**Solution:** The possible factors of 60 are 1, 2, 3, 4, 5, 6, 10, 12, 15, 20, 30 and 60.

$a = \alpha + \beta + \gamma$ = sum of roots = coefficient of $x^2$ = −2
$c = \alpha\beta\gamma$ = Product of three roots = 60

On mere inspection, we see-

$\alpha + \beta + \gamma$ = sum of roots = $-3 - 4 + 5 = -2$

$\alpha\beta\gamma$ = Product of three roots = $-3 \times -4 \times 5 = 60$

Hence, the factors are -

$$x^3 - 2x^2 - 23x + 60 = (x - 3)(x - 4)(x + 5)$$

**Verification:**

$b = \alpha\beta + \beta\gamma + \alpha\gamma$ = sum of the product of two = coefficient of x

$= (-3) \times (-4) + (-4) \times 5 + (-3 \times 5) = -23$

**Example:** Factorize $x^3 - 7x + 6$

**Solution:** This polynomial involves no term of $x^2$, so we re-write the equation in the standard format as discussed above.

$$x^3 - 7x + 6 = x^3 + 0x^2 - 7x + 6$$

Now the factors of 6 are 1, 2, 3 and 6. On inspection, we can locate the exact value of $\alpha$, $\beta$ and $\gamma$ that suits the value of a, b and c.

$a = \alpha + \beta + \gamma$ = sum of roots = coefficient of $x^2 = 0 = -1$
$-2 + 3$

$c = \alpha\beta\gamma$ = Product of three roots = $-1 \times -2 \times 3 = 6$

Hence, $x^3 - 7x + 6 = (x - 1)(x - 2)(x + 3)$

**Verification:**

$b = \alpha\beta + \beta\gamma + \alpha\gamma$ = sum of product of two = coefficient of x
$= -1 \times (-2) + (-2) \times 3 + (-1) \times 3$
$= 2 - 6 - 3$
$= -7$

Hence, result verified.

The above example illustrates that the Vedic Method to solve the cubic factor is not only easy to understand but a time saving technique because it involves no tedious or lengthy calculation, but merely inspection, which helps you to reach the result in no time. Always bear in mind, however, that this is valid as long as the cubic polynomial is factorizable.

# 16

# Quadratic Equations

## Introduction

An equation in the form of $ax^2 + bx + c = 0$ is called quadratic equation where $a \neq 0$. The Fundamental theorem of Algebra says that – *Every polynomial of m degree has m roots*. Since the degree of quadratic equation is 2, it will have two roots. The quadratic equation $ax^2 + bx + c = 0$ is solved by the formula –

$$x = \frac{-b \pm \sqrt{b^2 - 4ac}}{2a}$$

where $D = b^2 - 4ac$, is called the Discriminant. This Discriminant also helps us to decide the nature of roots of the equation.

The solution of Quadratic equations is a part of secondary school syllabus of every secondary board in India and this is not all, even in almost 90 % of competitive examination, 1– 2 questions based on quadratic equations are asked. The traditional method works well, but it takes a lot of time. In some special cases, the solution is arrived at in 5–10 minutes, whereas the Vedic method instantly gives you the answer in 5 –10 seconds and that too without the help of pen and paper. I have decided to leave the simple type of quadratic equation for readers as they can solve it by any of the following methods taught in Secondary Curriculum. The method described in CBSE/ ICSE syllabus of class X are--

a) Mid Term factorization
b) Completing the Square method
c) Quadratic formula

# Vedic method to solve Quadratic Equation

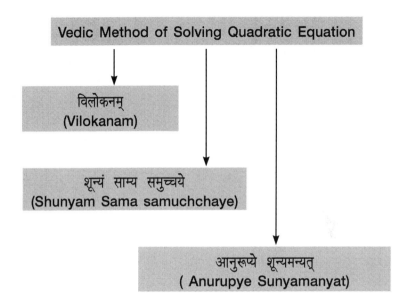

**Meaning of Vedic Sutras:**

1. Vilokanam (विलोकनम्): The literal meaning of this Vedic Sutra is Merely by Observation.
2. Sunyam Samya Samuccaye (शून्यं साम्य समुच्चये): This sutra is applicable to a large number of different cases. It literally means –when the samuchchaya is the same, equal it to zero. Samuchchaya is a technical term which has several meanings under different contexts; and we shall explain them, one at a time.
3. Anurupye Sunyamanyat (आनुरूप्ये शून्यमन्यत्): This sutra is useful in finding one root of a quadratic equation of special type. The literal meaning is – **If one is in ratio, the other one is zero.** It has many other applications in Vedic Mathematics

## Vilokanam (विलोकनम्) Sutra

We usually come across such equations in examination and the traditional method takes almost 5 minutes to solve, but the Vedic Method will let you solve such questions in less than a minute.

**Example**: $\dfrac{x + 2}{x + 1} + \dfrac{x + 1}{x + 2} = \dfrac{37}{6}$

**Traditional Method**:

Put $\dfrac{x + 2}{x + 1} = a$

Hence,

$$a + 1/a = 37/6$$

or, $\quad \dfrac{a^2 + 1}{a} = \dfrac{37}{6}$

or, $\quad 6\,a^2 + 6 = 37\,a$

or, $\quad 6\,a^2 - 37\,a + 6 = 0$

or, $\quad 6\,a^2 - 36\,a - a + 6 = 0$

or, $\quad 6\,a(a - 6) - 1\,(a - 6) = 0$

or, $\quad (6\,a - 1)\,(\,a - 6) = 0$

or, $\quad a = 6$ or $1/6$

Now,

$$\dfrac{x + 2}{x + 1} = 6 \qquad \text{or,} \qquad \dfrac{x + 2}{x + 1} = \dfrac{1}{6}$$

$\Rightarrow 6x + 6 = x + 2 \qquad \Rightarrow \quad 6x + 12 = x + 1$

$\Rightarrow 6x - x = 2 - 6 \qquad \Rightarrow \quad 6x - x = 1 - 12$

$\Rightarrow 5x = - 4 \qquad\qquad \Rightarrow \quad 5x = - 11$

$\Rightarrow x = - 4/5 \qquad\qquad \Rightarrow \quad x = - 11/5$

## Vedic Method:

Look at the LHS; you observe that the LHS is the sum of two reciprocals. The Vilokanam Vedic Sutra simply tells us to break the

RHS in such a way that they become the sum of two reciprocals and equate the RHS with any part of LHS. Let us understand the modus operandi with these examples.

**Example:** Solve, $\dfrac{x + 2}{x + 1} + \dfrac{x + 1}{x + 2} = \dfrac{37}{6}$

**Solution:** The LHS is the sum of two reciprocals, so we break the RHS into two such fractions so that it also becomes the sum of two reciprocal fractions.

$$\frac{x + 2}{x + 1} + \frac{x + 1}{x + 2} = \frac{37}{6} = 6 + \frac{1}{6}$$

Now, equate either of the LHS parts to both the term of RHS and solve them to find the value of x.

$$\frac{x + 2}{x + 1} = 6 \qquad \text{or, } \frac{x + 2}{x + 1} = \frac{1}{6}$$

$$\Rightarrow 6x + 6 = x + 2 \qquad \Rightarrow 6x + 12 = x + 1$$

$$\Rightarrow 6x - x = 2 - 6 \qquad \Rightarrow 6x - x = 1 - 12$$

$$\Rightarrow 5x = -4 \qquad \Rightarrow 5x = -11$$

$$\Rightarrow x = -4/5 \qquad \Rightarrow x = -11/5$$

**Example:** Solve, $\dfrac{x + 2}{x + 3} - \dfrac{x + 3}{x + 2} = \dfrac{15}{4}$

**Solution:** The LHS is the difference of two reciprocals, so we break the RHS into two such fractions so that it also becomes the difference of two reciprocal fractions.

$$\frac{x + 2}{x + 3} - \frac{x + 3}{x + 2} = \frac{15}{4} = 4 - \frac{1}{4}$$

Now, equate either of the LHS part to both the term of the RHS and solve them to find the value of x.

$$\frac{x + 2}{x + 3} = 4$$

$$\Rightarrow 4x + 12 = x + 2$$

$$\Rightarrow 4x - x = 2 - 12$$

$$\Rightarrow 3x = -10$$

$$\Rightarrow x = -10/3$$

or, $\dfrac{x + 2}{x + 3} = \dfrac{1}{4}$

$$\Rightarrow 4x + 8 = x + 3$$

$$\Rightarrow 4x - x = 3 - 8$$

$$\Rightarrow 3x = -5$$

$$\Rightarrow x = -5/3$$

**Example**: Solve, $\dfrac{x}{x + 1} + \dfrac{x + 1}{x} = \dfrac{169}{60}$

**Solution**: The LHS is the sum of two reciprocals, so we break the RHS in two such fractions so that it also becomes the sum of two reciprocal fractions. At first look, the RHS doesn't seem to be the sum of any such fraction. Now split the denominator in two parts.

$$60 = 2 \times 30$$
$$= 3 \times 20$$
$$= 4 \times 15$$
$$= 5 \times 12$$
$$= 6 \times 10$$

Now, find the sum of the squares of these factors and check when the sum is 169, the value of the numerator.

$$2^2 + 30^2 > 169$$
$$3^2 + 20^2 > 169$$
$$4^2 + 15^2 > 169$$
$$5^2 + 12^2 = 169 \text{ (Numerator)}$$

Hence, $169/60 = 12/5 + 5/12$

$$\frac{x}{x + 1} + \frac{x + 1}{x} = \frac{169}{60} = \frac{12}{5} + \frac{5}{12}$$

Now, equate either of the LHS parts to both the term of the RHS and solve them to find the value of x.

$$\frac{x}{x+1} = \frac{12}{5}$$

$$\Rightarrow 12x + 12 = 5x$$

$$\Rightarrow 12x - 5x = -12$$

$$\Rightarrow 7x = -12$$

$$\Rightarrow x = -12/7$$

or, $\dfrac{x}{x+1} = \dfrac{5}{12}$

$$\Rightarrow 12x = 5x + 5$$

$$\Rightarrow 12x - 5x = 5$$

$$\Rightarrow 7x = 5$$

$$\Rightarrow x = 5/7$$

**Example:** Solve, $\dfrac{3x+7}{2x-9} - \dfrac{2x-9}{3x+7} = \dfrac{56}{45}$

**Solution:** The LHS is the difference of two reciprocals, so we brea the RHS into two such fractions so that it also becomes the difference of two reciprocal fractions. At first look, the RHS doesn't seem to be the sum of any such fraction. Now split the denominator in two parts.

$$45 = 3 \times 15$$
$$= 9 \times 5$$

Now, find the difference of squares of these factors and check when the difference is found to be 56, the value of the numerator.

$$15^2 - 3^2 > 56$$
$$9^2 - 5^2 = 56 \text{ (Numerator)}$$

Hence, $\dfrac{56}{45} = \dfrac{9}{5} - \dfrac{5}{9}$

The question now becomes–

$$\frac{3x+7}{2x-9} - \frac{2x-9}{3x+7} = \frac{56}{45} = \frac{9}{5} - \frac{5}{9}$$

Now, equate either of the LHS parts to both the terms of the RHS and solve them to find the value of x.

$$\frac{3x + 7}{2x - 9} = \frac{9}{5}$$

$$\Rightarrow 15x + 35 = 18x - 81$$
$$\Rightarrow 15x - 18x = -81 - 35$$
$$\Rightarrow -3x = -116$$
$$\Rightarrow x = 116/3$$

or, $$\frac{3x + 7}{2x - 9} = \frac{5}{9}$$

$$\Rightarrow 27x + 63 = 10x - 45$$
$$\Rightarrow 27x - 10x = -45 - 63$$
$$\Rightarrow 17x = -108$$
$$\Rightarrow x = -108/17$$

## Sunyam Anyat (शून्यम् अन्यत) and Sunyam sama samuchaye (शून्यम साम्य समुच्चये)

Look at the following examples:

a) $$\frac{2}{x+2} + \frac{3}{x+3} = \frac{4}{x+4} + \frac{1}{x+1}$$

b) $$\frac{a+b}{x+a+b} + \frac{b+c}{x+b+c} = \frac{2b}{x+2b} + \frac{a+c}{x+a+c}$$

Minute observation proves the above equation to be a kind of Quadratic Equation, so it will certainly have two roots. The above type of equation can be solved by using two Vedic Sutras.

Let us take each example and solve it by using Vedic Sutras.

$$\frac{2}{x+2} + \frac{3}{x+3} = \frac{4}{x+4} + \frac{1}{x+1}$$

Now let us inspect it by using **Sunyam Anyat** Vedic sutra which was earlier used in solving a special type of simultaneous equation.

The formula says – if one is in ratio, the other one is zero.
Ratio of the constant term in LHS = 2/2 + 3/3 = 1 + 1 = 2
Ratio of the constant term in RHS = 4/4 + 1/1 = 1 + 1 = 2
Since the ratios of the constant term in both sides are equal, by Sunyam Anyat sutra x = 0.

Now what about the second root?

The second root will be extracted by using another Vedic Sutra called **Sunyam Sam samucchaye**. The sutra says – if sum of numerator on both the sides are the same, equate the sum of the denominator equal to zero.

$$N_1 + N_2 \text{ in LHS} = 2 + 3 = 5$$
$$N_1 + N_2 \text{ in RHS} = 4 + 1 = 5$$

Since they are equal, we have to equate the sum of the denominator equal to zero.

$$D_1 + D_2 = 0$$
$$\Rightarrow x + 2 + x + 3 = 0$$
$$\Rightarrow 2x + 5 = 0$$
$$\Rightarrow x = -5/2$$

Therefore, two roots are $x = 0$ and $x = -5/2$

**Example 2:** $\dfrac{a + b}{x + a + b} + \dfrac{b + c}{x + b + c} = \dfrac{2b}{x + 2b} + \dfrac{a + c}{x + a + c}$

**Solution:** Ratio of constant term in LHS =

$$\dfrac{a + b}{a + b} + \dfrac{b + c}{b + c} = 1 + 1 = 2$$

Ratio of constant term in RHS =

$$\dfrac{2b}{2b} + \dfrac{a + c}{a + c} = 1 + 1 = 2$$

Since the ratio of LHS and RHS are the same, $x = 0$ (By Sunyam Anyat Vedic Sutra). The second root will be extracted using another Vedic Sutra – Sunyam Sam samuccaye.

$$N_1 + N_2 \text{ in LHS} = a + b + b + c = a + 2b + c$$
$$N_1 + N_2 \text{ in RHS} = 2b + a + c = a + 2b + c$$

Since they are equal, we have to equate the sum of the denominator equal to zero.

$$D_1 + D_2 = 0$$
$$\Rightarrow x + a + b + x + b + c = 0$$
$$\Rightarrow 2x + a + 2b + c = 0$$
$$\Rightarrow x = -(a + 2b + c)/2$$

Hence, two roots are– $x = 0$ and $x = -(a + 2b + c)/2$

## Sunyam Sam Samuchchaye (शून्यम साम्य समुच्चये)

The third case is an example of simple quadratic equation. The traditional method is cumbersome but the Vedic Method is a one-line solution and does not even require pen and paper. You will be amazed to see how effective Vedic method is in such a special type of Quadratic equation. It is well known that a quadratic equation has two roots. This special type of equation can be identified by summing up the numerators and denominators of both the sides. If they are ound to be equal, then Sunyam Sam Samuchchaye Vedic sutra will be applicable in such a case.

$N_1 + N_2 = D_1 + D_2 = 0$, gives the first root of the equation

And, $\quad N_1 - D_1 = N_2 - D_2 = 0$, gives the second root of the equation

**Example**: Solve $\quad \dfrac{3x + 4}{6x + 7} = \dfrac{5x + 6}{2x + 3}$

Solution: The minute observation of the question helps us to identify that this question falls in a special category of quadratic equations, where the sum of the numerator and denominator is found to be equal to $8x + 10$.

**For the first root**

$$N_1 + N_2 = D_1 + D_2 = 0$$
$$N_1 + N_2 = 3x+4 + 5x + 6 = 8x + 10$$
$$D_1 + D_2 = 6x + 7 + 2x + 3 = 8x + 10$$

$\Rightarrow 8x + 10 = 0$

or, $x = -5/4$

**For the second root**

$$N_1 - D_1 = N_2 - D_2 = 0$$
$$\Rightarrow N_1 - D_1 = 3x + 4 - 6x - 7 = 0$$
$$\Rightarrow -3x - 3 = 0$$
$$\Rightarrow x = 1$$

or, $N_2 - D_2 = 5x + 6 - 2x - 3 = 0$

$\Rightarrow 3x + 3 = 0$

$\Rightarrow x = -1$

Hence, the two roots of the above equation are $x = -5/4$ and -1.

**Example:** Solve $\dfrac{3x + 6}{6x + 3} = \dfrac{5x + 4}{2x + 7}$

**Solution:** On simple observation, it is evident that the above equation is a quadratic equation and here the sum of numerator and denominator is found to be equal to $8x + 10$. We now apply the Sunyam sam Samuchchaya sutra to obtain the two roots of the quadratic equation.

**For the first root**

$$N_1 + N_2 = D_1 + D_2 = 0$$
$$N_1 + N_2 = 3x + 6 + 5x + 4 = 8x + 10$$
$$D_1 + D_2 = 6x + 3 + 2x + 7 = 8x + 10$$

$\Rightarrow 8x + 10 = 0$

$\Rightarrow$ or, $x = -5/4$

**For the second root**

$$N_1 - D_1 = N_2 - D_2 = 0$$
$\Rightarrow N_1 - D_1 = 3x + 6 - 6x - 3 = 0$
$\Rightarrow -3x + 3 = 0$
$\Rightarrow x = 1$

or, $N_2 - D_2 = 5x + 4 - 2x - 7 = 0$

$\Rightarrow 3x - 3 = 0$

$\Rightarrow x = 1$

Hence the two roots of the above equation are $x = -5/4$ and 1.

# Casting Out Nines

## Introduction

Nine is a wonderful number in mathematics. It is the largest one digit number in the Hindu-Arabic system. Besides that, it has many other features too–

- It is the least number that can be expressed as the sum of cubes of the first two numbers $1^3 + 2^3 = 9$.
- 9 is the maximum number of cubes that one needs to sum to any positive integer.
- If the sum of the digits of a number is the multiple of nine, it is divisible by nine.
  Example: 34521426 is divisible by 9 because the sum of digits $(3 + 4 + 5 + 2 + 1 + 4 + 2 + 6 = 27)$ is divisible by 9.
- Multiply any number by 9 – however big it is, the sum of the digits of the number will always result in 9.
  Example: a) $123 \times 9 = 1107$
  Sum of the digit $= 1 + 1 + 0 + 7 = 9$
  b) $459873 \times 9 = 4138857$
  Sum of the digit $= 4 + 1 + 3 + 8 + 8 + 5 + 7 = 36$
  $= 3 + 6 = 9$
- See another pattern of 9
  $1/9 = 0.1111....$   $2/9 = 0.222.....$   $3/9 = 0.3333.....$
  $4/9 = 0.4444.....$   $5/9 = 0.555.....$   $6/9 = 0.6666....$
  $7/9 = 0.7777.....$   $8/9 = 0.888.....$

Casting out nines is one of the easiest mathematical methods to check the accuracy of arithmetical operations like – Addition, Subtraction, Multiplication, Division, Square, Cube etc. It is a way to quickly check your mathematical operation, which will tell you that you have got a sum wrong, or that it is probably right. This method is also known as Chinese Remainder Theorem. In Vedic Mathematics, it is popularly known as नवशेष

## Fundamental rules of Vedic Sutra Navasesh (नवशेष):

**Casting out Nines** literally means to throw nines. Now let us focus on its working.

- Add the digits of a number across, dropping out 9, to get a single figure. If it is not a single figure, add the digits obtained so as to get a single figure less than 9.
- 9 is not taken into account in this process, as a digit sum of 9 is the same as a digit sum of zero.
- If the number is made up of all 9's and/or all sub-additions of 9's, then its number digit is zero.
- In subtraction, while applying this rule for verification, you may encounter a negative number. Add 9 to the negative number to make it positive. Example – if your answer is – 5 , convert it to positive integer by adding 9, i.e. – 5 = – 5 + 9 = 4

## How does this method work?

**Example:** Find the digit sum of 4954653

**Solution:**

Digit sum of 4954653= 4+ 9+ 5 + 4 + 6 + 5 +3 = 36

Since 36 is a double figure number, to get a single figure, we have to add it again.

Digit sum of 36 = 3+ 6 = 9 =0

(As the rule discussed above says that the digit sum of 9 is the same as the digit sum of zero)

The digit sum of 4954653 can be done in other way very easily. As discussed above, we need not take 9 into account.

Digit sum of 4954653 = 4 + 9̶ + 5 + 4 + 6 + 5 + 3

The two group of number 5+ 4 and 6 +3 can easily be left out while finding the digit sum of 4954653, as their sum is equal to 9.

**Example**: Find the digit sum of 4379348568219

**Solution**: Add all the digits

$$4 \; 3 \; 7 \; 9̶ \; 3 \; 4 \; 8 \; 5 \; 6 \; 8 \; 2 \; 1 \; 9̶$$

The four groups of numbers have the digit sum 9 which need not taken into account.

Hence the final sum = digits left out

= 4 + 3 + 8 = 15

Digit sum of 15 = 1 + 5 = 6

**Example**: Find the digit sum of 2345689

**Solution**: Add the digits, removing the pairs whose sum is 9.

$$2 \qquad 3 \qquad 4 \qquad 5 \qquad 6 \qquad 8 \qquad 9̶$$

**Checking for Addition**

**Rule**:

Find the digit sum (DS) of all numbers to be added individually and also the digit sum (DS) of the result/ answer you get at the end. If the digit sum (DS) for both is found to be equal, then it is time to cheer as your calculations are right!

**Example:** Verify 347385 + 569384 + 258769+ 381730 + 429464
= 1986732

**Verification:**

|  | Digit sum of number |
|---|---|
| 347385 | 3 |
| 569384 | 8 |
| 258769 | 1 |
| 381730 | 4 |
| + 429464 | 2 |
| 1986732 | ? |

LH.S. = Digit sum of 1986732 = 1+9+8+6+7+3+2 = 0
R.H.S. = Sum of the Digit sum of number = 3+8+1+4+2 = 0
Since LHS = RHS
Result Verified

**Example:** Verify 5087643 + 8432397+ 3854009 + 2197565 =
19571614

**Verification:**

|  | Digit sum of number |
|---|---|
| 5087643 | 6 |
| 8432397 | 0 |
| 3854009 | 2 |
| + 2197565 | 8 |
| 19571614 | ? |

LH.S. = Digit sum of 19571614 = 1+9+5+7+1+6+1+4 = 7
R.H.S. = Sum of the Digit sum of number = 6+0+2+8 = 7
Result verified

**Example:** 10045 + 34567 + 88888 + 234 = 145734

**Verification:**

|  | Digit Sum of number |
|---|---|
| 10045 | 1 |
| 34567 | 7 |
| 88888 | 4 |
| + 234 | 0 |
| 145734 | ? |

LH.S. = Digit sum of 145734 = 1 + 4 + 5 + 7 + 3 + 4 = 6
R.H.S. = Sum of the Digit sum of number = 1 + 7 + 4 + 0 = 3
Since, LHS ≠ RHS
Result Incorrect

## Check for Subtraction

**Rule:**

The process of subtraction is the same as applied in addition. Remember that the value of the digit sum of minuend should be greater than that of the subtrahend. In case the digit sum of the minuend is less than that of the subtrahend, don't panic and do the simple arithmetic calculation and the answer in this case will be negative one. Add 9 to the result to make it positive.

**Example:** Verify 42587 − 35769 = 6818

| **Verification:** | Digit Sum of number |
|---|---|
| 42587 | 8 |
| − 35769 | 3 |
| 6818 | ? |

LHS = Digit sum of 6818 = 5
RHS = 8 − 3 = 5
Since LHS = RHS

Hence Result Verified

**Example:** Verify 3456928734 −1958762087 = 1498166647

| **Verification:** | Digit Sum of number |
|---|---|
| 3456928734 | 6 |
| −1958762087 | 8 |
| 1498166647 | ? |

LHS = Digit sum of 1498166647 = 1+4+9+8+1+6+6+6+4+7
=7
RHS = 6 − 8 = −2

Since, the result is negative,add 9 to it to make it positive. Hence,
RHS = −2 + 9 = 7
LHS = RHS
Hence, result verified

**Example**: Verify 502568 − 369876 = 138692

**Verification**:

| | Digit Sum of number |
|---|---|
| 502568 | 8 |
| 369876 | −3 |
| 138692 | ? |

LHS = Digit sum of 138692 = 1 + 3 + 8 + 6 + 9 + 2 = 2
RHS = 8 − 3 = 5
Since, LHS ≠ RHS
Hence result is incorrect.

## Check for Multiplication

Multiplication is one of the most error-prone areas amongst the eight fundamental arithmetical operations. Students spend much of their precious time in checking and rechecking the result as they are not sure about getting the result 100 % correct. The casting out nines method will undoubtedly boost your level of confidence and prove a panacea. Once you are good enough to apply this method, you will love to see the tremendous change in your calculating ability as you are now equipped to fight the menace of calculation errors. In a fraction of a minute, you would be able to check whether you are wrong or right. Let us enjoy this beauty in Multiplication with the help of some examples.

**Rule**:

We know that Multiplicand X Multiplier = Result. First, find the digital roots of Multiplicand, multiplier and product. If the result in LHS and RHS comes out to be equal, then you are correct, else you need to multiply it again.

**Example**: Verify; 3456 x 9999 = 34556544

**Solution**: Digital sum of Multiplicand = 3 + 4 + 5 + 6 = 0
Digital sum of Multiplier = 9 + 9 + 9 + 9 = 0
Digital sum of Result = 3 + 4 + 5 + 5 + 6 + 5 + 4 + 4 = 0

LHS = DS of Multiplicand x DS of Multiplier = 0 x 0 = 0
RHS = DS of Result = 0
LHS =RHS
Result verified

**Example**: Verify, 8 5 9 4 2 x 3 0 5 4 = 2 6 2 4 6 0 8 6 8

**Solution**: Digital sum of Multiplicand = 8 + 5 + 9 + 4 + 2 = 1
Digital sum of Multiplier = 3 + 0 + 5 + 4 = 3
Digital sum of Result = 2 + 6 + 2 + 4 + 6 + 0 + 8 + 6 + 8 = 6

LHS = DS of Multiplicand x DS of Multiplier = 1 x 3 = 3
RHS = DS of Result = 6
LHS ≠ RHS
Result Incorrect

**Example**: Verify, 28762 x 65490 = 1883623380

**Solution**: Digital sum of Multiplicand = 2 + 8 + 7 + 6 + 2 = 7
Digital sum of Multiplier = 6 +5 + 4 + 9 + 0 = 6
Digital sum of Result = 1 + 8 + 8 + 3 + 6 + 2 + 3 + 3 +8 + 0 = 6

LHS = DS of Multiplicand x DS of Multiplier = 7 x 6 = 42 = 4 + 2 = 6
RHS = DS of Result = 6
LHS = RHS
Result verified

## Check for Division

The casting out nines method described in the very beginning will suffice to check the division operation effectively.

**Rule:**

We know that **Dividend = Divisor x Quotient + Remainder**

To verify whether the Quotient and Remainder obtained on dividing a certain number by another number is correct or not, find the digit sum of Dividend, Divisor, Quotient and Remainder and put the value of digits sum in LHS and RHS. If the same digit sum is obtained on both the sides, it ultimately tells you that you have performed the right operation. Let us take some examples to understand how effectively this method works for division.

**Example:** Verify 876543 ÷ 123, Q = 7126 and R = 45

**Verification:**

> Here, Dividend = 876543
> Digit sum of Dividend = 8 + 7 + 6 + 5 + 4 + 3 = 6
> Divisor = 123
> Digit sum of Divisor = 1 + 2 + 3 = 6
> Quotient = 7126
> Digit sum of quotient = 7 + 1 + 2 + 6= 7
> Remainder = 45
> Digit sum of Remainder = 4 + 5 = 0

Putting the digit sum value in the given formulae, we get,

L.H.S = Digit sum of Dividend = 6
R.H.S = Divisor x Quotient + Remainder
      = 6 x 7 + 0 = 42
Digit sum of 42 = 6

Hence LHS = RHS
Result Verified

**Example**: Verify 287695 ÷ 3456, Q = 83 and R = 847

**Verification**:

      Here, Dividend = 287695

      Digit sum of Dividend = 2 + 8 + 7 +6 + 9 + 5 = 1

      Divisor = 3456

      Digit sum of Divisor = 3 + 4 + 5 + 6 = 0

      Quotient = 83

      Digit sum of quotient = 8 + 3 = 2

      Remainder = 847

      Digit sum of Remainder = 8 + 4 + 7 = 1

Putting the digit sum value in the given formulae, we get,

L.H.S = Digit sum of Dividend = 1

R.H.S = Divisor x Quotient + Remainder

      = 0 x 2 + 1 = 1

Hence LHS = RHS

Result Verified

**Example**: Verify 8240376287 ÷ 58769, Q = 140216 and R = 22183

**Verification**:

      Here Dividend = 8240376287

      Digit sum of Dividend = 8 + 2 + 4 + 0 + 3+7 + 6 + 2 + 8 + 7 = 2

      Divisor = 58769

      Digit sum of Divisor = 5 + 8 + 7 + 6 + 9 = 8

      Quotient = 140216

      Digit sum of Quotient = 1 + 4 + 0 + 2 + 1 + 6= 5

      Remainder = 22183

      Digit sum of Remainder = 2 + 2 + 1 + 8 + 3 = 7

Putting the digit sum value in the given formulae, we get,

L.H.S = Digit sum of Dividend = 2

R.H.S = Divisor x Quotient + Remainder

      = 8 x 5 + 7 = 47

Digit sum of 47 = 2
Hence LHS = RHS
Result Verified

## Check for Square

**Rule:**

Square of any number can be verified by means of the Casting out Nines technique very easily. It is a proven fact that the square of a number is the multiplication of the same number by itself, so the process of multiplication can also be used to check the square of a number. For fast calculation, readers are advised to go through this chart and if possible, keep the chart in mind.

| Number | 1 | 2 | 3 | 4 | 5 | 6 | 7 | 8 | 9 |
|--------|---|---|---|----|----|----|----|----|----|
| Square | 1 | 4 | 9 | 16 | 25 | 36 | 49 | 64 | 81 |
| DS | 1 | 4 | 0 | 7 | 7 | 0 | 4 | 1 | 0 |

**Example:** Verify $(42)^2 = 1764$

**Verification:**

LHS = Digital sum of $(42)^2 = (4 + 2)^2 = 0$
RHS = Digital sum of $1764 = 1 + 7 + 6 + 4 = 0$
LHS = RHS
Result verified

**Example:** Verify $(106)^2 = 12336$

**Verification:**

LHS = Digital sum of $(106)^2 = (1 + 0 + 6)^2 = 4$
RHS = Digital sum of $12336 = 1 + 2 + 3 + 3 + 6 = 7$
LHS $\neq$ RHS
Result Incorrect

**Example**: Verify $(938)^2 = 879834$

**Verification**:

LHS = Digital sum of $(938)^2 = (9 + 3 + 8)^2 = 4$
RHS = Digital sum of $879834 = 8 + 7 + 9 + 8 + 3 + 4 = 3$
LHS $\neq$ RHS
Result Incorrect.

**Example**: Verify $(2345)^2 = 5499025$

**Verification**:

LHS = Digital sum of $(2345)^2 = (2 + 3 + 4 + 5)^2 = 7$
RHS = Digital sum of $5499025 = 5 + 4 + 9 + 9 + 0 + 2 + 5 = 7$
LHS = RHS
Result verified

## Check for Cube

**Rule**:

Find the DS of the cube of the number and check whether the result obtained after cubing the number is the same as the result you have extracted. Always remember that the final Digit Sum of the LHS and RHS should be of one digit. Moreover, the combination of 9 and multiples of 9 should be discarded at very first sight, so that the journey to reach the final digit sum becomes easier. For the convenience of readers, the digit sum of cubes from 1 to 9 is given here.

| Number | 1 | 2 | 3 | 4 | 5 | 6 | 7 | 8 | 9 |
|--------|---|---|---|---|---|---|---|---|---|
| Cube | 1 | 8 | 27 | 64 | 125 | 216 | 343 | 512 | 729 |
| Digit Sum | 1 | 8 | 0 | 1 | 8 | 0 | 1 | 8 | 0 |

Let us understand the concept with some examples.

**Example:** Verify $(13)^3 = 2197$

**Verification:**
LHS = Digital sum of $(13)^3 = (1 + 3)^3 = 1$
RHS = Digital sum of 2197 = 2 + 1 + 9 + 7 = 1
LHS = RHS
Result verified

**Example:** Verify $(96)^3 = 884736$

**Verification:**

LHS = Digital sum of $(96)^3 = (9 + 6)^3 = 0$
RHS = Digital sum of 884736 = 8 + 8 + 4 +7 +3 +6 = 0
LHS = RHS
Result verified

**Example:** Verify $(928)^3 = 799168552$

**Verification:**

LHS = Digital sum of $(928)^2 = (9 + 2 + 8)^3 = (10)^3 = 1$
RHS = Digital sum of 799168552 = 7 + 9 + 9 + 1 + 6 +8
+5 + 5 + 2 = 7
LHS $\neq$ RHS
Result Incorrect

# 18

# Trigonometry

## Introduction

The word Trigonometry deals with the ratio of sides in a right-angled triangle. The basic concepts of trigonometry which are taught today in our school curriculum are to find–

- At secondary level – Sin, cos(ine), tan(gent), cot(angent), sec(ant) and cosec(ant)
- However, at the senior secondary level, we learn to find the value of multiple and sub-multiple angle and in doing this, we are forced to memorize scores of formula.

Let me remind you of the following concepts of trigonometry, which you have learnt at the secondary level. In a right-angled triangle ABC with angle C = $90^0$, the various trigonometric ratios are defined as follows:

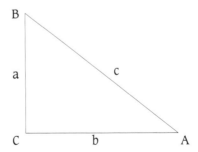

| TRIGONOMETRIC RATIO | VALUE | VEDIC TRIPLET CASE |
|---|---|---|
| Sin A | a / c | 1st value / 3rd value |
| Cos A | b/c | 2nd value / 3rd value |
| Tan A | a/ b | 1st value / 2nd value |
| Cot A | b /a | 2nd value / 1st value |
| Sec A | c / b | 3rd value / 2nd value |
| Cosec A | c / a | 3rd value / 1st value |

In the above table, the side a, b and c are taken as triplet. You must be remembering the famous Pythagoras theorem which states – 'In a right angle triangle the square of hypotenuse is equal to the sum of square of other two sides.'

Hypotenuse $^2$ = Perpendicular $^2$ + Base$^2$

$c^2 = a^2 + b^2$

It means all the three sides are related to a special condition. In order to get faster calculation in Vedic Mathematics, you need to memorise the Pythagorean triplet. Here is a table to help you.

| a | b | c |
|---|---|---|
| 3 | 4 | 5 |
| 5 | 12 | 13 |
| 6 | 8 | 10 |
| 7 | 24 | 25 |
| 8 | 15 | 17 |
| 9 | 40 | 41 |
| 9 | 12 | 15 |
| 10 | 24 | 26 |
| 11 | 60 | 61 |
| 12 | 15 | 18 |
| 3n | 4n | 5n |

Let us understand the benefit of this table. Suppose in a right angle triangle, two of its legs are given as 3 and 4 and you need to find its largest leg i.e. Hypotenuse. Traditionally, what will you do? You will use Pythagoras' theorem to find the third side. On the other hand, if you are familiar with the above triplets, then in a fraction of a second you can say that the length of the third side is 5.

## How to compute Pythagorean Triplet

*Case 1: When one of the numbers is odd.*

We know that the square of an odd number is odd. This square is the sum of two consecutive numbers. If you know one value of the triplet the other two can be easily traced.

See the following example and guess what I am trying to say–

$$3^2 = 9 = 4 + 5$$
$$5^2 = 25 = 12 + 13$$
$$9^2 = 81 = 40 + 41$$

The above examples show that when we break the square of an odd number into two consecutive numbers, they form a triplet. (3, 4, 5) , (5, 12, 13), (9, 40, 41)... are the triplet.

Now the question is –How do we find the consecutive pairs of numbers? The answer is very simple. Divide the square of a number by 2 and round it off.

In the first example, the square of 3 is 9, i.e. $3^2 = 9$. On dividing it by 2, we get 4.5. When 4.5 is rounded off to its next higher and lower numbers, we get 5 and 4. Thus (3– 4 – 5) form a triplet where the square of the highest number is the sum of the square of other two numbers.

In the second example, the square of 5 is 25. When 25 is divided by 2, we get 12.5. The next lower and higher round off number to 12.5 is obviously 12 and 13. Interestingly, 5–12–13 forms a triplet.

The third example has been left for the reader.

*Case 2: When one of the numbers is even.*

**Rule:** Divide the number by 2, 4, 8 etc. to get an odd number. Once you get the odd number, follow the rule explained above in case 1. If you have divided the even number by 2, 4, 8, etc., the final answer will have to be multiplied by the same number to get the triplet.

**Example:** One value of the triplet is 6, find the other two.

**Solution:** Divide 6 by 2 to get an odd number. The triplet as explained above for 3 is 4 and 5. Since we have divided the number by 2, we have to multiply all the values by 2 to form the triplet. Hence our triplet containing 6 is – (6 – 8 – 10)

## Computing Trigonometric Ratio

Suppose you are given the following example to solve.

**Example:** If Tan A = 8 /15, find the value of other trigonometric ratios?

The traditional method will use different formula to arrive at the result.

Cot A = 1 / tan A = $^{15}/_8$
Sec$^2$A = 1 + tan$^2$A = 1 + $^{64}/_{225}$ = $^{289}/_{225}$, so Sec A = $^{17}/_{15}$
Cos A = 1 / sec A = 15 /17
Sin$^2$A = 1 – cos$^2$A = 1 – 225 /289 = 64 / 289 so, Sin A = 8 / 17
Cosec A = 1/ sin A = 17 / 8.

Now let us see how the triplet helps us to compute the values of other trigonometric ratios. From the above table on triplets, you can find t if–

a = 8          b = 15          then     c = 17.

Now move to the trigonometric table and find the value of other trigonometric ratios in no time.

Sin A = a / c = 8 / 17
Cos A = b / c = 15 / 17

The other ratios can be easily found as cosec A is the inverse of sin A and sec A is the inverse of cos A.
Take another example.

**Example**: If cosec A = 61/11 find tan A?

By Pythagorean triplet, the third number in the series of 11, 61 is obviously 60.
Since, sin A = 1/ cosec A = 11/ 61 = first value / third value

i.e. a = 11, c = 61 then b = 60

Now from the trigonometric table given above,

Tan A = first value/ second value = a/ b = 11 / 60

The above example are enough to prove that the triplet method of Vedic mathematics is interesting, easy to understand and time saving. Now, let us extend the value to twice the angle.

## Computing trigonometric ratio of twice the angle (2A)

As discussed above, the triplet for angle A is – a, b and c. On extending the result for twice the angle i.e. 2A, we can find the triplets for the angle 2A are $2ab$, $b^2 - a^2$ and $c^2$.

Take an example.

If sin A = 3/5, find tan 2A?

Traditional method–

We have $\cos^2 A = 1 - \sin^2 A$
$$= 1 - 9 /25 = 16 / 25$$
$$\cos A = 4 / 5.$$
Moreover, tan A = sin A / cos A.
$$\Rightarrow \tan A = 3 / 4.$$

Hence tan $2A = 2 \tan A / 1 + \tan^2 A$

$$= \frac{2 \times \frac{3}{4}}{1 - (3/4)^2} = 24/7$$

Now let's view the problem by the Vedic triplet method.

We have, sin A = 3/5

Here a = 3 and c = 5, so obviously b = 4 (see triplet table).

Now make the triplet for twice the angle, i.e. for 2 A.

| The triplets are– | 2ab, | $b^2 - a^2$ and | $c^2$ |
|---|---|---|---|
| | 2 x 3 x 4 | $4^2 - 3^2$ | $5^2$ |
| | 24 | 7 | 25 |

Hence,

Tan 2 A = $1^{st}$ value / $2^{nd}$ value = 24 / 7.

**Example:** if cos A = 9 / 41, find cos 2A

**Solution:** First find the triplet for angle A.

Cos A = $2^{nd}$ value/ $3^{rd}$ value = b / c

Hence, the missing triplet a = 40.

So we have, a = 40, b = 9 and c = 41.

Now the triplets for 2A are–

| 2ab | $b^2 - a^2$ and | $c^2$ |
|---|---|---|
| 2 x 40 x9 | $40^2 - 9^2$ | $41^2$ |
| 720 | 1519 | 1681 |

So,

Cos2A = $2^{nd}$ value / $3^{rd}$ value = 1519 /1681

## Computing trigonometric relation for thrice the angle (3 A)

We have so far seen the triplets for angle A and 2A – let's extend it for 3A to find the value of Sin 3A, cos 3A and tan 3A.

The triplets for 3A is–

| $3ac^2 - 4a^3$ | $4b^3 - 3bc^2$ | and | $c^3$ |
|---|---|---|---|

Let me make it clear with an example.

**Example:** If tan A = 7/ 24 find sin 3A and cos 3A ?

**Solution:** The triplet for angle A –

Tan A = 7/24 = 1$^{st}$ value / 2$^{nd}$ value

Hence the missing part of the triplet
$\qquad$ = 25 = 3$^{rd}$ value, i.e. a = 7, b = 24 and c = 25
Let us find the value of triplet for angle 3A.
The triplets are

| $3ac^2 - 4\,a^3$ | $4b^3 - 3\,bc^2$ and | $c^3$ |
|---|---|---|
| 3 x 7 x 25$^2$ – 4 x 7$^3$ | 4 x 24$^3$ – 3 x 24 x 25$^2$ | 25$^3$ |
| 11753 | 10296 | 15625 |

(For square and cube of a number refer the respective chapter)
Hence sin 3A = 1$^{st}$ value/ 3$^{rd}$ value = 11753 / 15625
$\qquad$ Cos 3A = 2$^{nd}$ value/ 3$^{rd}$ value = 10296 / 15625

## Computing trigonometric relation for half the angle(A/2)

If the triplet for the angle A is a b and c then the triplets for A/2 is–

$\qquad$ a $\qquad$ b + c $\quad$ and $\sqrt{(b + c)^2 + a^2}$

**Example:** If sin A = 12/13 find the value of tan A/2.

**Solution:** We have, sin A = 1$^{st}$ value / 3$^{rd}$ value

Here, a = 12 $\quad$ c = 13 $\quad$ therefore from the triplet table $\quad$ b = 5.

Now find the triplet for A/2.

| a | b + c | and $\sqrt{(b + c)^2 + a^2}$ |
|---|---|---|
| 12 | 5 + 13 | $\sqrt{18^2 + 12^2}$ |
| 12 | 18 | $\sqrt{468}$ |
| 12 | 18 | $6\sqrt{13}$ |

Hence tan A/2 = 1$^{st}$ value/ 2$^{nd}$ value = 12/18 = $^2/_3$

Now let us summarise the triplet in the given table.

| Angle | a | b | c |
|-------|-----|----------|------------------------|
| A | a | b | c |
| 2A | 2 ab | $b^2 - a^2$ | $c^2$ |
| 3A | $3ac^2 - 4 a^3$ | $4b^3 - 3 bc^2$ | $c^3$ |
| A/2. | a | b+c | $\sqrt{(b + c)^2 + a^2}$ |

The above table will help you immensely to find the different trigonometric ratios with ease and thus save your precious time. Once you are well equipped with the method of finding square and cube you can find the calculation involved in sin 3A, cos 2A etc quite easily. Now let us extend the concept of triplets in finding the value of compound angles.

**Computing Trigonometric ratio of (A+ B):**

If you are given the trigonometric ratio of two of the angles, then using the concept of triplets you can find the ratio of the sum of these angles.

Suppose you are given sin A = 3/5 and sin B = 8/17 and asked to find the value of sin (A +B) or cos (A+B), you will have to apply the following formula in order to find the value of compound angle A+ B of different trigonometric ratios–

Sin (A +B) = sin A · cos B + cos A · sin B

cos (A + B ) = cos A · cos B − sin A · sin B

But before that, you have to find cos A and cos B and put these values in the desired formula. Now let us see how the Vedic method helps you to find the value of compound angle.

If the triplets of angle A is x, y, z and triplet for B is X, Y and Z, then the triplet for the angle A+B is given by:

| A | $x$ | y | z |
|---|---|---|---|
| B | X | Y | Z |
| A+ B | yX + x Y | yY – xX | z Z |

**Example:** If sin A = 3/5 and sin B = 8/17, then find the value of Sin (A+B) and cos (A+B)

**Solution:** Let us first draw the triplet table for the angle A and B

| A | 3 | 4 | 5 |
|---|---|---|---|
| B | 8 | 15 | 17 |
| A + B | 3 x 15 + 4 x 8 | 4 x 15 – 3 x 85 | x 17 |
| | = 77 | = 36 | = 85 |

Hence, sin (A+ B) = 77/85

Cos (A + B) = 36 /85

**Example:** If sin A = 7/25 and sin B = 8/17 then find the value of Sin (A+B) and cos (A+B)

Solution: Let us first draw the triplet table for the angle A and B

| A | 7 | 24 | 25 |
|---|---|---|---|
| B | 8 | 15 | 17 |
| A + B | 7 x 15 + 24 x 8 | 24 x 15 – 7 x 8 | 25 x 17 |
| | = 297 | = 304 | = 425 |

Hence, sin (A+ B) = 297/425

Cos (A + B) = 304 /425

## Computing Trigonometric ratio of A – B

Computing trigonometric ratio of difference of angle can be computed with the help of triplet in a few seconds and that is the beauty of the Vedic method of calculation. I am not describing here the traditional method used in our classroom. Let us first learn how to compute the triplets fro difference of angles.

If the triplets of angle A is x, y, z and triplet for B is X, Y and Z then the triplet for the angle A+B is given by:

| A | x | y | z |
|---|---|---|---|
| B | X | Y | Z |
| A − B | xY −Xy | xX + yY | z Z |

**Example:** If sin A = 7/25 and sin B = 8/17 then find the value of Sin (A −B) and cos (A−B)

**Solution:** Let us first draw the triplet table for the angle A and B

| A | 7 | 24 | 25 |
|---|---|---|---|
| B | 8 | 15 | 17 |
| A − B | 7 x 15 −24 x 8 | 24 x 15 + 7 x 8 | 25 x 17 |
| | = −77 | = 416 | = 425 |

Hence, sin (A − B) = −77/425

Cos (A − B) = 416 /425

Example: If sin A = 3/5 and sin B = 8/17 then find the value of Sin (A−B) and cos (A−B)?

**Solution:** Let us first draw the triplet table for the angle A and B

| A | 3 | 4 | 5 |
|---|---|---|---|
| B | 8 | 15 | 17 |
| A − B | 3 x 15 − 4 x 8 | 4 x 15 + 3 x 8 | 5 x 17 |
| | = 13 | = 84 | = 85 |

Hence, sin (A− B) = 13/85

Cos (A − B) = 84 /85

Hope you have enjoyed the journey of Trigonometry using the triplet method and the traditional method. A little practice and memorization of the triplet will ease the process of calculation almost one tenth that of the traditional method.

# Questions for Practice

## 1. Addition

Add the following by using the appropriate method–

1. 75934 + 87628 + 34879 + 14093 + 256
2. 876549762 + 345982769 + 470154897 + 284579657 + 145469885
3. 762874 +3476928 + 593487 + 8752546 + 274039
4. 4876 + 8752684 + 187049 + 48998
5. 6938789159689 + 5248792300000 + 7895248301554 + 8736200145932 + 5260148530489

| 6. | Km | m | 7. | Kg | g |
|---|---|---|---|---|---|
| | 256 | 145 | | 87 | 958 |
| | 253 | 874 | | 11 | 025 |
| | 326 | 099 | | 05 | 652 |
| + | 2 | 450 | + | 1 | 006 |

| 8. | Rs | P | 9. | Rs | P |
|---|---|---|---|---|---|
| | 325 | 25 | | 659 | 38 |
| | 625 | 32 | | 968 | 15 |
| | 325 | 62 | | 159 | 64 |
| | 2 | 06 | | 6 | 98 |
| + | 28 | 65 | + | 98 | 36 |

## 2. Subtraction

Subtract the following

1)　　6 6 6
　　− 4 8 2

2)　　6 9 8 0
　　− 5 7 9 8

3)　　9 8 6 7 9 8 5 4 6
　　− 6 8 9 7 9 7 9 7 9

4)　　8 7 0 5 6 9 7 6 9
　　− 8 2 4 8 9 3 9 8 7

5)　　4 7 9 8 6 7
　　− 3 6 4 7 8 0

6)　　7 5 6 2 7 9
　　− 2 9 8 7 0 5

7)　　9 2 6 7 9 9 8 2 5
　　− 5 4 9 8 9 9 9 9 9

8)　　4 6 8 7 9 3
　　− 3 5 9 7 0 2

9)　　4 0 0 0 0 0 0 0
　　− 7 8 8 9 6 3 2

10)　　7 0 0 0 0 0 0 0 0
　　　　　− 8 8 8 8 8 8

## 3. Multiplication

Multiply the following

a)　36 x 34
b)　87 x 83
c)　128 x 122
d)　112 x 998
e)　688 x 988
f)　107 x 95
g)　9997 x 9998
h)　252 x 248
i)　148 x 149
j)　506 x 494
k)　2487 x 9999
l)　87904 x 99999
m)　8284 x 99
n)　43427 x 9999
o)　144x 9999
p)　279 x 331
q)　7628 x 4287
r)　144 x 66
s)　248 x 128
t)　82765 x 42897
u)　5628 x 3047
v)　983 x 994 x 1005
w)　1003 x 1007 x 1009
x)　876 x 602
y)　56 x 66 x 65 x 64
z)　992 x 994 x 996 x 998

## 4. Multiply through Observation

a)　8247 x 11
b)　24876 x 11
c)　987204 x 11
d)　576 3 x 11
e)　24076 x 111
f)　4213x 1111
g)　24076 x 11
h)　4217 x 111
i)　928764 x 25

j)   428 x 25               k)  24376 x 25            l)   4876 x 50
m)  7654 x 50             n)  82765 x 125          o)  87 x 125
p)   34 x 51               q)  89 x 51               r)   82 x 51
s)   765 x 75              t)  654 x 51               u)  98764 x 75
v)   982 x 1111           w)  87 x 83               x)  76 x 25
y)   843 x 51             z)  9652 x 50

## 5. Multiplication in Algebra

a)   $(2x + 3y) \times (4x + 7y)$
b)   $(a + 7z) \times (2a + 11z)$
c)   $(x^2 + 4z) \times (9x^2 + 7z)$
d)   $(2x^2 + 4x + 7) \times (x^2 + 7x - 9)$
e)   $(5x^2 - 9x - 8) \times (4x^2 - 7x + 8)$
f)   $(7x^2 - 6x) \times (x^2 - 3x + 4)$
g)   $(12x^2 - 7x) \times (3x + 4)$
h)   $(8x^2 + 4) \times (7x^2 + 2x + 5)$
i)   $(4x + 7y) \times (2x + 6y)$
j)   $(2a + 4c) \times (a + c)$

## 6. Division

Divide by using the appropriate method

a)   $4532 \div 854$        b) $2101532 \div 879$     c) $12345 \div 8888$
d)   $11203 \div 8999$      e) $2100012 \div 8997$    f) $2002002 \div 89997$
g)   $13999 \div 112$       h) $11329 \div 1132$      i) $239479 \div 11203$
j)   $39999 \div 1812$      k) $2112 \div 87$         l) $46823564 \div 3542$
m)  $5362968527 \div 9213649875$
n)   $111038972168 \div 895475582$

## 7. Square

Find the square of the following by using the appropriate method.

| a) 15 | b) 25 | c) 35 | d) 65 | e) 85 | f) 95 |
|-------|-------|-------|-------|-------|-------|
| g) 105 | h) 115 | i) 125 | j) 44 | k) 76 | l) 54 |
| m) 29 | n) 168 | o) 185 | p) 97 | q) 99 | r) 109 |
| s) 168 | t) 2356 | u) 6254 | v) 2591 | | |

## 8. Square Root

Find the square root of the following by using the appropriate Method.

| a) 4473225 | b) 10329796 | c) 14047504 | d) 2116 |
|------------|-------------|-------------|---------|
| e) 4225 | f) 6889 | g) 59049 | h) 125316 |
| i) 169744 | j) 1265625 | k) 99920016 | l) 931225 |
| m) 45369 | n) 790321 | o) 78978769 | p) 45319824 |
| q) 69906321 | r) 41254929 | | |

## 9. Square Root of Irrational

Find the square root of the following

| a) 37 | b) 164 | c) 175 | d) 43 | e) 231 | f) 198 |
|-------|--------|--------|-------|--------|--------|
| g) 69 | h) 287 | i) 631 | j) 259 | k) 599 | l) 1221 |

## 10. Cube

Find the cube of the following by using the appropriate Vedic Sutra.

| a) 13 | b) 19 | c) 25 | d) 36 | e) 46 | f) 54 | g) 69 |
|-------|-------|-------|-------|-------|-------|-------|
| h) 87 | i) 104 | j) 113 | k) 208 | l) 315 | m) 86 | n) 97 |
| o) 98 | p) 16 | q) 94 | r) 96 | s) 97 | t) 102 | |

## 11. Cube Root

Find the cube root of the following numbers.

a) 6892          b) 636056          c) 314432
d) 8365427       e) 1061208         f) 8489664
g) 143055667     h) 9800344         i) 8615125
j) 33076161      k) 2196            l) 83453453
m) 143877824     n) 830584          o) 300763

## 12. Fourth Power

Find the fourth power of the following

a) 23    b) 17    c) 14    d) 98    e) 42
f) 21    g) 24    h) 73    i) 76    j) 55

## 13. Fourth Roots of a Number

Find the fourth root of the following numbers

a) 50625         b) 14776336        c) 1048576
d) 20736         e) 4879681         f) 279841
g) 83521         h) 37015056

## 14. Simultaneous Equation

Solve for x and y

a)   23x + 29y = 42; 46x+ 14y = 84
b)   7y – 2x = 5; 8y + 7x = 15
c)   30u + 44 v = 10; 40u + 55 v = 13
d)   152x – 378y = –74; –378x + 152y = – 604
e)   x + 3y = 6; 2x – 3y = 12
f)   x + y = 9; 8x – y = 0
g)   217x + 131y = 913; 131x + 217y = 827

h) $2x + 5y = 13; 2x + 3y = 4$
i) $5x + 3y = 19xy; 7x - 2x = 8xy$
j) $x + y = 63; 3x - 4y = 0$

## 15. Cubic Factorization Factorize

a) $x^3 + 13x^2 + 31x - 45$
b) $x^3 - 2x^2 - x + 2$
c) $x^3 - 3x^2 - 9x - 5$
d) $y^3 - 2y^2 - 29y - 42$
e) $x^3 - 10x^2 - 53x - 42$
f) $x^3 - 23x^2 + 142x - 120$
g) $y^3 - 7y + 6$

## 16. Quadratic Equation

Solve for x

a) $(x - 2/x + 2)^2 + 6 = 5 (x - 2 / x + 2)$
b) $(7x - 1 / x)^2 + 3(7x - 1/x) = 18$
c) $6(y - 3 / 2y + 1) + 1 = 5(y - 3/2y + 1)^2$
d) $1/x - 4 - 1 / x - 7 = 11/30$
e) $x / x + 1 + x + 1/x = 13/6$
f) $3/x - 1 - 1/x - 2 - 1/x - 3 = 0$
g) $2x - 3 / x - 1 - 4( x - 1/ 2x - 3) = 3$
h) $x/ x + 1 + x + 1/x = 25/12$
i) $2x + 3 / x + 4(x / 2x + 3 ) = 13/3$
j) $(2x + 1) + 3 / 2x + 1 = 4$

## 17. Casting out Nines

Verify the following result

a) $112065 + 360085 + 289872 + 156345 = 918367$
b) $4998 + 6789 + 5715 + 4837 + 8976 = 31315$

c)  7534 + 2459 + 1932 + 6547 = 16472

d)  37467 + 35647 + 285 +10085 =82876

e)  3746735 − 2837546 = 909189

f)  876542 − 32548 − 698547= 145447

g)  658723 + 154639 − 369847 +367 = 443882

h)  588 X 512 = 301056

i)  842 X 858 = 722536

j)  966 X 973 = 939918

k)  13579 ÷ 975, Q = 13, R = 904

l)  7238761 ÷524, Q=13184, R=225

m)  11199171 ÷99979, Q= 112, R = 1523

n)  87265 X 32117 = 2802690005

o)  6471 X 6212 = 40197852

p)  $(207)^2$ = 42849

q)  $(2134)^2$ = 4553856

r)  $(3247)^2$ = 10542169

s)  $(12)^3$ = 1729

t)  $(65)^3$ = 98002

## 18. Trigonometry

a) If sin A = 8/17, find the value of other five trigonometric ratio.

b) If tan A= 9/40, find the value of sin2A, cos2A, tan2A and sec2A

c) If cos A = 3/5, find the value of sin3A, cos3A and tan3A

d) If sin A = 8/10, find the value of sin A/2, cos A/2 and tan A/2

e) If sin A = 3/5, sin B = 8/17 find sin (A+B), cos (A+B)

f) If sin A = 12/13, sin B = 8/17 find sin (A − B), cos (A− B)

## Answer

### Chapter 1: Addition

1. 212790      2. 212276970      3. 13859874

4. 8993607      5. 34079178437664      6. 838.568 Km

7. 105.641Kg   8. Rs 1306.90      9. Rs 1892.51

## Chapter 2: Subtraction

1. 184  2. 1182      3. 297000567  4. 45675782
5. 115087          6. 457574    7. 376899826
8. 109091          9. 32110368  10. 699111112

## Chapter 3: Multiplication

a) 1224       b) 7221        c) 15616     d) 111776
e) 679744     f) 10165       g) 99950006  h) 62496
i) 22052      j)249964       k) 24867513  l) 8790372096
m) 820116     n) 424226573   o) 1439856   p) 92349
q) 2701236    r) 13824       s) 31744     t) 3550370205
u) 17148516   v) 981987510   w) 1019111189
x) 527352     y) 15375360    z) 980139600384

## Chapter 4: Multiplication through observation

a)  90717     b)  273636     c)  10859244  d)  63393
e)  2672436   f)  4680643    g)  264836    h)  468087
i)  23219100  j)  10700      k)  609400    l)  243800
m)  382700    n)  10345625   o)  10875     p)  1734
q)  4539      r)  4182       s) 57375      t)  33354
u)  7407300   v)  1091002    w)  7221      x)  1900
y)  42933     z)  482600

## Chapter 5: Multiplication in Algebra

a)  $8x^2 + 26xy + 21y^2$
b)  $2a^2 + 25az + 77z^2$
c)  $9x^4 + 43 x^2z + 28z^2$
d)  $2x^4 + 18x^3 + 17x^2 + 13x - 63$
e)  $20x^4 - 71x^3 + 71x^2 - 135x - 64$
f)  $7x^4 + 27x^3 + 46x^2 - 24x$
g)  $36x^3 + 27x^2 - 28x$
h)  $56x^4 + 16x^3 + 68x^2 + 8x + 26$
i)  $8x^2 + 38xy + 42x^2$
j)  $2a^2 + 6ac + 4c^2$

# Chapter 6: Division

Quotient Remainder

a)  5  262
b)  2390  722
c)  1  3457
d)  1  2204
e)  233  3711
f)  22  22068
g)  124  111
h)  10  9
i)  21  4216
j)  22  135
k)  24  24
l)  13219  1866
m) 5  8207
n)  124  0

# Chapter 7: Square

a) 225      b) 625      c)1225      d) 4225     e)7225     f) 9025
g) 11025    h) 13225    i) 15625    j) 1936     k)5776     l) 2916
m) 841      n) 28224    o) 34225    p) 9409     q)9801     r) 11881
s) 28224    t) 5550736  u) 39112516 v) 6713281

# Chapter 8: Square Root

a) 2115     b) 3214     c) 3748     d) 46       e) 65      f) 83
g) 243      h) 354      i) 412      j) 1125     k) 9956    l) 965
m) 213      n) 889      o) 8887     p) 6732     q) 8361    r) 6423

# Chapter 9: Square root of Irrational

a) 6.08     b) 12.8     c) 13.23    d) 6.57     e) 15.7    f) 14.07
g) 8.06     h) 16.94    i) 25.12    j) 16.09    k) 24.47   l) 34.94

## Chapter 10: Cube

a) 2197    b) 6859    c) 15625   d) 46656    e) 97336
f) 157464   g) 328509   h) 658503   i) 1124864   j) 1442897
k) 8998912   l) 31255875   m) 636056   n) 912673   o) 941192
p) 4096    q) 830584    r) 884736   s) 912673   t) 1061208

## Chapter 11: Cube Root

a) 18    b) 86    c) 68    d) 203    e) 102    f) 204
g) 523   h) 214   i) 205   j) 321    k) 21    l) 437
m) 524   n) 94   o) 67

## Chapter 12: Fourth Power

a) 279841   b) 83521   c) 38416    d) 92236816   e) 3111696
f) 194481   g) 331776   h) 28398241   i) 33362176   j) 9150625

## Chapter 13: Fourth Root

a) 15   b) 62   c) 32   d) 12   e) 47   f) 23   g) 17   h) 78

## Chapter 14: Simultaneous Equation

a)   $x = 42/13, y = 0$
b)   $x = 1; y = 1$
c)   $u = 1/5; v = \frac{1}{4}$
d)   $x = 2; y = 1$
e)   $x = 6; y = 0$
f)   $x = 1; y = 8$
g)   $x = 3; y = 2$
h)   $x = -19/4; y = 9/2$
i)   $x = 1/3; y = \frac{1}{2}$
j)   $x = 36; y = 27$

## Chapter 15: Cubic Factorization

a)   $(x - 1) (x + 5) (x + 9)$
b)   $(x - 2) (x - 1) (x + 1)$

c) $(x - 1) ( x + 1)(x - 5)$

d) $(y+2) (y+3) (y - 7)$

e) $(x + 1) ( x +3) (x - 14)$

f) $(x - 1) ( x - 10 )(x - 12)$

g) $(y - 1) (y +3)(y - 2)$

## Chapter 16: Quadratic Equation

a) $x = -6, -4$

b) $x = \frac{1}{4}, 1/13$

c) $x = 4, 13/2$

d) $x = 3, -1/2$

e) $x = 2; y = -3$

f) $x = 4 \pm \sqrt{3}$

g) $x = \frac{1}{2}, 4/3$

h) $x = 3, -4$

i) $x = 3, -9/2$

j) $x = 0, 1$

## Chapter 17: Casting Out Nines

| | | | |
|---|---|---|---|
| a) Correct | b) Correct | c) Incorrect | d) Incorrect |
| e) Correct | f) Correct | g) Correct | h) Correct |
| i) Incorrect | j) Correct | k) Correct | l) Incorrect |
| m) Correct | n) Correct | o) Correct | p) Correct |
| q) Incorrect | r) Incorrect | | |

## Chapter 18: Trigonometry

a) Cos A = 15/17, Tan A = 8/15 etc.

b) Sin 2A = 720/1681, cos 2A = 1519/1681; tan 2A = 720/1519

c) Sin 3A = 236/125; cos 3A = −117 / 125; tan 3A = −236/117

d) Sin A/2 = $\sqrt{2}$ /$\sqrt{5}$; cos A/2 = 2/$\sqrt{5}$; tan A/2 = 1/$\sqrt{2}$

e) Sin (A+B) = 77/85; cos (A+B) = 36/85

f) Sin (A − B) = 140/221; cos (A − B ) = 171/221

# Feedback from Students

1. Vedic mathematics reduces problem-solving time by almost one–tenth. A wonderful method to enjoy mathematical calculations.
   —Monika Chauhan, Kendriya Vidyalaya Tiruananthpuram

2. Easy-to-learn technique. It has revolutionized mathematical concepts.
   —Rajeev Ranjan, BCA

3. Vedic mathematics has helped me to improve my calculating ability. I enjoy calculating even bigger sums in seconds.
   —Soumya Vishwakarma, Class 10, St. Xavier's School Raj Niwas Marg, Delhi.

4. I attended the Vedic maths class by Mr Rajesh Thakur and learnt the art of fast calculations. It is a superb and mind-blowing technique. It has increased the speed of my calculations.
   —Shivam Maggo, Class 11, Montfort School, Ashok Vihar

5. A unique method to solve multiplication, square, square root and cube root. Superb technique for speedy calculations.
   — Pawan Bharti, Ganga International School, New Delhi

6. An ancient method of solving modern mathematics. Wonderful method to boost my calculating ability.
   —Pooja Singh, DAV School, Rohini

7. I met Mr Rajesh Thakur when he came for a lecture on Vedic mathematics to our Innovative School in Rajkot (Gujarat). His method of teaching Vedic mathematics is superb. I have witnessed the change in the mindset of my students and seen their improvement in calculations.

—Dr. Atul K Vyas, Director, Innovative School, Rajkot

8. Fantastic, superb, awesome, mind blowing—and many more such words describe the beauty of Vedic mathematics.

—Riya Chanda, Hakim Para Senior
Secondary Balika Vidyalaya, Siliguri

9. Vedic mathematics is a superb method of intelligently guessing answers without using pen and paper. It has given me the potential to do my arithmetic calculations faster than my friends.

—Simran Ghosh, Ram Krisha Sardamani Vidyapeeth
(Senior Secondary), Ashram Para, Siliguri

10. I first learnt the art of faster calculation from Mr Rajesh Thakur when I was in Kendriya Vidyalaya, AFS, Suratgarh. This technique has changed my calculating ability and speed, and now I can boast of being able to multiply a 5-digit multiplication in my head and that too in 20 seconds!

—Satyam Thakur, Kendriya Vidyalaya, Pune

11. Mathematical calculations now seem easier and enjoyable. I could qualify for the MBA test due to my understanding of Vedic mathematics

—Er. Akshay Bahal, Shalimar Bagh

12. An ancient method to encounter present day mathematical calculations in smarter ways, and that too mentally, in no time at all. I have witnessed the change in my calculating speed, and my success in the MBA entrance is due to the technique of Vedic mathematics.

—Nilesh Dey, Manager, LG company, Noida.

13. A few months ago, I joined the Vedic maths class and learnt the art of faster calculations. Besides calculation, Vedic mathematics also teaches the unique method of checking the calculations—and this helps me a lot in the subject of accounting.

   —Megha Agarwal, Queen Mary Schoo, 1 Model Town, Delhi

14. Vedic mathematics has changed my calculating ability tremendously.

   —Anmol Chadda, Heritage School, Rohini.

15. Vedic mathematics is a must-learn-tool for every student. Sixteen years ago, Rajesh Thakur had visited St Xavier's school, Sahebganj to teach the Vedic technique of calculation to class 9-10 students, when he himself was a student! His innovative and simple way of teaching mathematics made a long-lasting impression on my mind.

   —D. Srivastava, Sr Teacher, St Xavier School, Sahebganj

16. Mr Rajesh Thakur has made an outstanding impact through his mathematical tricks on the minds of students. I have seen an enormous change in the attitude of students towards mathematics, after attending the Vedic maths seminar.

   —Mr. K A Sorte, Principal, GBSSS Nithari

17. I got the chance to observe Mr Rajesh Thakur conducting a Vedic mathematics class for teachers in Palanpur, Gujrat. That's when I saw the strength of Vedic mathematics. It is a powerful technique to perform swift mental calculations and I suggest that every student be trained in this technique.

   —Dr Deepak Sinha, Principal, DAVPS, NTPC Dadri

# Bibliography

1. Krisna, Tirtha Ji, Swami Bharti. *Vedic Mathematics*, Delhi: Motilal Banarsi Das Publishers, 1965.
2. Rao, Balachandra, S. *Indian Mathematics and Astronomy*, Bangalore: Jnana Deep Publication, 1996.
3. Topic on Vedic mathematics taken from Mathematics Course book, Rajasthan Board, (Class IX and X), 2007.
4. Muthy, Bhanu, T.S. *A Modern Introduction to Ancient Indian Mathematics*, New Age International Publisher,1992.
5. Williams, Kenneth. *Triples*, New Delhi: Motilal Banarsidass, 2003.
6. Gupta, Atul. *The Power of Vedic Mathematics*, New Delhi: Jaico Books, 2004.
7. Singhal, Vandana. *Vedic Mathematics for all ages*, Motilal Banarsidass Publishers Pvt. Ltd., 2008.
8. Bathia, Dhaval. *Vedic Mathematics Made Easy*, Jaico Books, 2010.
9. Thakur, Rajesh Kumar. *Mathematical Magic (With Vedic Sutra)*, Rising Publishers, 2007.
10. Sikri, Ashima. *Vedic Mathematics for Entrance Exam*, Oswal Books, 2009.
11. Handley, Bill. *Speed Mathematics*, John Wiley & Sons, Inc., 2003.
12. Glover, James. *Vedic Mathematics for School (Part 1, 2 and 3)*, Motilal Banarsidass,1995.
13. Cutler, Ann and Mc Shane, Rudolph. *The Trachtenberg Speed System of Basic Arithmetic*, Rupa Publications, 1985.

14. वैदिक गणित विहंगम दृष्टि ;भाग ९, २, व ३द्ध.. डा. कैलाश विश्वकर्मा . शिक्षा संस्कृति उत्थान न्यासए 2010.

15. Singhal, Aditi. *How to Become a Human Calculator*, S.Chand Publishing, 2011.

16. Tyra, M. *Magical Book on Quicker Maths*, Banking Service Chronicle Publication, 1999.

17. Thakur, Rajesh Kumar. *Vedic Mathematics*, Unicorn Books, 2009.

18. Bhushan, S; Gupta, B.S. *Calculation without Tears*, Ocean Books Pvt. Ltd., 2010.

19. Naseer, Vali. *Speed Mathematics Using the Vedic System*, Lulu Press, 2004.

20. Kapoor, S.K. Dr. *Glimpses of Vedic Mathematics*, New Delhi: Arya Book Depot, 2003.

21. Nicholas, A.P; Williams, K; Pickles, J. *Vertically and Crosswise*, New Delhi, 2010.

22. Devi, Shakuntala. *Figuring: The Joy of Numbers*, New Delhi: Orient Paperbacks,1986.

23. Kumar, Pradeep. *Vedic Mathematics*, New Delhi: Sterling Publishers Pvt. Ltd., 2002.

24. Sastri, Ramanand Pt. *Vedic Mathematics*, New Delhi: Arihant Publisher , 2011.

25. Benjamin, Arthur; Shermer, Michael; Nye, Bill. *Secret of Mental Maths*, New Delhi: Crown Publishing Group, 2006.

26. Kapoor, S.K. Dr. *Vedic Maths for All*, New Delhi: Lotus Press, 2011.

27. Unkalkar, V.G. *Magical World of Mathematics (Vedic Maths)*, Bangalore: Vandana Publisher, 2005.

28. Vadhanam, Ganita; Rao, M. Seeta Rama, ISERVE, Hyderabad, 2009.

29. Unkalkar, V.G. *Excel with Vedic Mathematics*, Bangalore: Vandana Publisher, 2008.

30. Prakash, Vijay. *Essence of Vedic Mathematics*, Vasan Publication, 2008.

31. Kapoor, S.K. Dr. *Vedic Maths for All*, New Delhi: Lotus Press, 2008.
33. Chauthaiwale, Shriram M; Kolluru, Ramesh. *Enjoy Vedic Mathematics*, Sri Sri Publication Trust, 2010.

Printed in Great Britain
by Amazon